Water and Health

Water and Health

Edited by

Alpana Kateja

RAWAT PUBLICATIONS

Jaipur • New Delhi • Bangalore • Guwahati • Kolkata

ISBN 978-81-316-0871-5

Published by
Prem Rawat for **Rawat Publications**
Satyam Apts, Sector 3, Jawahar Nagar, Jaipur 302 004 (India)
Phone: 0141 265 1748 / 7006 Fax: 0141 265 1748
E-mail: info@rawatbooks.com
Website: www.rawatbooks.com

New Delhi Office
4858/24, Ansari Road, Daryaganj, New Delhi 110 002
Phone: 011 2326 3290

Also at *Bangalore, Guwahati* and *Kolkata*

Printed at Nice Printing Press, New Delhi

Contents

Preface

Water is the most vital resource for human life as is also evident from the fact that river basins have cradled almost all civilizations of the world. Water is truly a unique gift of nature to mankind. It is essential for sustaining all forms of life, food production, economic development and for general well-being. It is impossible to substitute water, difficult to purify once polluted and expensive to transport.

Growing water scarcity along with deteriorating quality is one of the most noticeable phenomena of the present century. With surging economic activities, vertically and horizontally, both quantity and quality are decreasing resulting in major economic, social and environmental costs. Rapid increase in population, urbanization, changing life style have left per capita water availability to less than 2000 cubic metres per year and the actual usable quantity is merely 1000 cubic metres per year. The compound annual growth rate of demand for water driven by rapid growth in agricultural and municipal demand is one of the highest (2.8%) in the world. Only 13 per cent of the sewage is treated and an estimated 700 million Indians have no access to a proper toilet, making India's sanitation execrable. Deterioration in the quality and quantity of both surface and ground water has been very well articulated in a number of studies.

Water is imperative for all forms of life on this planet. However, this life giving substance, when contaminated, causes outbreak of diseases and destruction of eco-system. Water has very strong influence on human health – both as a means to reduce disease and as a medium through which disease causing agents may be transmitted. The adverse impact of water on health derives principally from the consumption of water containing pathogenic organisms or toxic chemicals and the use of inadequate volume of water that leads to poor hygiene.

In the above perspective, a modest attempt has been made by the editor to focus attention on the problem by incorporating literary studies of eminent scholars in this field. It is hoped that the volume will prove useful for planners and academicians.

The present volume on 'Water and Health' is an outcome of the National Seminar on 'Water Resource and Health Hazards' organized under the aegis of UPE 'Study of Ecology and Economy of Rajasthan', University of Rajasthan, Jaipur during 13-14 March 2015. The seminar was organized as part of the mandate of the UGC supported 'University with Potential for Excellence' (UPE). I place on record my gratitude for all the financial support which made possible the successful organization of the seminar. The two-day seminar intended to serve as a forum for all the stakeholders and succeeded to achieve the objectives of: examining the ground realities in the water sector; problems and challenges faced by people due to poor quality of water; and also to ponder upon mitigation measures. Organization of seminar was not possible without the support of University administration, colleagues, students and research scholars of the Department of Economics. I express special thanks to all of them.

I would like to convey my appreciation all the contributors. Without their cooperation and effort, outcome in the form of present volume would not have been possible. I have been fortunate in obtaining excellent support from academicians of varied backgrounds and discriplines who are united in the belief that human existence and the well-being are undisputedly supreme and must not, in any case, be compromised with.

My special thanks goes to all those who helped me put all contributions together. Thanks are also due to Rawat Publications, especially to Pranit ji for taking the project forward.

Jaipur **Alpana Kateja**

Acronyms

AARP	Annual Actual Rate of Progress
AAS	Atomic Absorption Spectrophotometer
ADI	Acceptable Daily Intake
ADP	Adenosine Diphosphate
AIDS	Acquired Immune Deficiency Syndrome
AOAC	Association of Official Analytical Chemists
ATP	Adenosine Triphosphate
ATSDR	Agency for Toxic Substances and Disease Registry
B. subtilis	*Bacillus subtilis*
BHC/HCH	Benzene Hexachloride or Hexachlorocyclohexane
BIS	Bureau of Indian Standard
BMI	Body Mass Index
BOD	Biochemical Oxygen Demand/Biological Oxygen Demand
BPA	Bisphenol A
BW	Body Weight
C	Centigrade
Ca	Calcium
Cd	Cadmium
CDC	Centre for Disease Control and Prevention
CDD	Cooling Degree-Days

Ce	Cerium
CE	Capillary Electrophoresis
CESCR	Committee on Economic, Social and Cultural Rights
CES-D	Centre for Epidemiologic Studies Depression
CETPs	Common Effluent Treatment Plants
CGWB	Central Ground Water Board
CNS	Central Nervous System
Co	Cobalt
COD	Chemical Oxygen Demand
CPCB	Central Pollution Control Board
Cr	Chromium
Cu	Copper
CVD	Cardio Vascular Diseases
CWD	Annual Maximum Number of Consecutive Wet Days
day^{-1}	Per Day
DDT	Diphenyl Trichloroethane
DIM	Daily Intake of Metals
DNA	Deoxyribonucleic Acid
E. coli	*Escherichia coli*
EC	Electrical Conductivity
EAG	Empowered Action Group
EC	European Commission
ECD/ED	Electron Capture Detector
ECETOC	European Chemical Industry Ecology and Toxicology Centre
EDCs	Endocrine Disrupting Compounds
eNOS	Endothelial Nitric Oxide Synthase
EPA	Environment Protection Agency
ERK5	Extracellular-Signal-Regulated Kinase 5
F	Fluoride
FAD	Flavin Adenine Dinucleotide
FAO	Food and Agriculture Organization
FDA	Food and Drug Administration
Fe^{+2}	Ferrous
Fe^{+3}	Ferric

FEF25	Forced Expiratory Flow at 25% of Expired Volume During FVC (Forced Vital Capacity) Test
FEF50	Forced Expiratory Flow at 50% of Expired Volume During FVC Test
FEV 1	Forced Expiratory Volume in the 1st second
FSH	Follicle Stimulating Hormone
G3PD	Glyceraldehyde 3-Phosphate Dehydrogenase
GC	Gas Chromatography
GDC	Ghaggar Diversion Channel
GDP	Gross Domestic Product
GI/GIT	Gastro Intestinal Tract
GNP	Gross National Product
GoI	Government of India
GSI	Geological Survey of India
GW	Groundwater
Hb	Hemoglobin
HCB	Hexachlorobenzene
HEPT	Heptachlor
HIV	Human Immunodeficiency Virus
HNO_3	Nitric Acid
HONO	Nitrous Acid
HRI	Health Risk Index
IARC	International Agency for Research on Cancer
IDDM	Insulin-Dependent Diabetes Mellitus
IGNP	Indira Gandhi Nahar Pariyojna
IMD	Indian Meteorological Department
IMR	Infant Mortality Rate
iNOS	Inducible Nitric Oxide Synthase
Interleukin-10	Human Cytokine Synthesis Inhibitory Factor (CSIF)
IPCC	Intergovernmental Panel on Climate Change
IPEN	International POPs Elimination Network
IPM	Integrated Pest Management
IQ	Intelligence Quotient
IRS	Indian Remote Sensing Satellites
IS	Indian Standard
JECFA	Joint Expert Committee on Food and Agriculture

K	Potassium
KMO	Kaiser-Meyer-Olkin
KOH	Potassium Hydroxide
LC	Liquid Chromatography
LDL	Low-density Lipoprotein
LH	Luteinizing Hormone
LISS	Linear Imaging Self Scanning Sensor
m/z	Mass/Charge
m^3	Cubic Metre
mbgl	Metres Below Ground Level
MCM	Million Cubic Metres
MDG	Millennium Development Goal
MDS	Myelodysplastic Syndrome
Mg	Magnesium
$MgAl\text{-}CO_3$	Magnesium Aluminium Carbonate
MIA	Mewar Industrial Area
MLD	Million Litres Per Day
MM	Multiple Myeloma
MPTP	1-Methyl-4-Phenyl-1,2,3,6-Tetrahydropyridine
MRL	Maximum Residue Limits
MRM	Multi Residue Method
MS	Mass Spectrometry
MT	Metric Ton
mV	Millivolts
N	Nitrogen
N_2O	Nitrous Oxide
N_2O_3	Dinitrogen Trioxide
NaCl	Sodium Chloride
NaCl	Sodium Chloride
NAD	Nicotinamide Adenine Dinucleotide
NADH	Nicotinamide Adenine Dinucleotide Dehydrogenase
NADPH	Nicotinamide Adenine Dinucleotide Phosphate Reduced
NAS	National Academy of Sciences
NGT	National Green Tribunal
NHL	Non-Hodgkins Lymphoma
NHS	National Network of Hydrograph Stations

Ni	Nickle
NO	Nitrogen Oxide
NO_2	Nitrogen Dioxide
NO^{2-}	Nitrite
NO_3/ NO_3^-	Nitrate
NOCs	N-nitroso Compounds
NOx	Mono-Nitrogen Oxides NO and NO_2 (Nitric Oxide and Nitrogen Dioxide)
NR	Nitrate Reductase
NRC	National Research Council
NTP	National Toxicology Program
O_3	Ozone
OBC	Other Backward Category
OC	Organochlorine
OCP	Organochlorine Pesticide
$ONOO^-$	Peroxynitrite
OPIDP	Organophosphate-Induced Delayed Polyneuropathy
OPs	Organophosphates
ORS	Oral Rehydration Solutions
OW	Observation Well
ox-LDL	Oxidized Low-density Lipoprotein
Oxy cyt b_5	Oxiginated Cytochrome b_5
p,p'DDD	Parapara Dichlorodiphenyl Ddichloro Ethane
p,p'DDE	Parapara Dichlorodiphenyl Trichloro Ethylene
p,p'DDT	Parapara Dichlorodiphenyl Trichloro Ethane
PAN	Pesticide Action Network
Pb	Lead
PBDE	Polybrominated Diphenyl Ethers
PCA	Principal Component Analysis
PCBs	Polychlorinated Biphenyls
PD	Parkinson's Disease
PDGF	Platelet-Derived Growth Factor
PET	Potential Evapotranspiration
PFA	Prevention of Food and Adulteration
PFOA or C8	Perflurorooctanoic Acids
PFOS	Perfluorooctane Sulfonate

pH	Potential of Hydrogen
PHED	Public Health Engineering Department
PNS	Peripheral Nervous System
POPs	Persistent Organic Pollutants
ppb	Parts Per Billion
ppbv	Parts Per Billion by Volume
ppm	Parts Per Million
PRCPTOT	Total Amount of Precipitation
PRECIS	Providing Regional Climates for Impact Studies
PVC	Poly Vinyl Chloride
PZ	Piezometres
R&D	Research and Design
RBC	Red Blood Cells
RCEP	Royal Commission on Environmental Pollution
RDA	Recommended Dietary Allowance
Red cyt b_5	Reductase Cytochrome b_5
R_{fd}	Reference Dose
RIVM	Rijksintituut voor de Volksgenzondheid en Milieuhygiëne
ROS	Reactive Oxygen Species
RRAP	Required Rate of Annual Progress
RTDC	Rajasthan Tourism Development Corporation
RX1	Annual Maximum 1-Day Precipitation
SC	Scheduled Castes
SCF	Scientific Committee for Food
SEER	Surveillance, Epidemiology and End Results
sGC	Soluble Guanylate Cyclase
SH	Slim Holes
SH	Sulfhydryl
SHS	Smoke Inhaled Unintentionally
SO_4	Sulphate
SRS	Sample Registration System
SSA	Superoxide Scavenger Activity
ST	Scheduled Tribes
Staph. albus	*Staphylococcus albus*
STPs	Sewerage Treatment Plants

STS	Soft Tissue Sarcomas
T_3	Triiodothyronine
T_4	Thyroxine
TB	Tuberculosis
TCM	Thousand Million Cubic Feet
TDS	Total Dissolved Solids
TF	Transfer Factor
Th1	T Helper 1 Cell
Th2	T Helper 2 Cell
TN	Minimum Temperature
TNT	Tri Nitro Toluene
TSC	Total Sanitation Campaign
TSH	Thyroid Stimulating Hormone
TSNA	Tobacco Specific Nitrosamine
TX	Maximum Temperature
U5MR	Under Five Mortality Rate
UNDP	United Nation Development Programme
UNICEF	United Nations International Children's Emergency Fund
USA/US	United States of America
USP DI	United States Pharmacopeia-Drug Information
VOCs	Volatile Organic Compounds
WCRF	World Cancer Research Fund
WHO	World Health Organization
WMO	World Meteorological Organisation
WPR	Work Participation Rate
Zn	Zinc
ã HCH	Lindane
ì	Micro or 10^{-6}
1,3-BPG	1,3-Bisphosphoglyceric Acid
^{15}N	Isotope of Nitrogen

1

Introduction

Mankind has always been fully aware of importance of water in sustaining life on the earth, which is evident from the fact that this life-giving element was an extraordinary and omnipresent element even in Rig Veda:

> *Ya aapo divyaa utavaa sravanti khanitrimaa utavaa yaa swayarn jaa |*
> *samudraartha yaa soochayapaavakaasta aap devi iha mamavanthu ||*
>
> (Rig-Veda, VII 49.2)
>
> *The water which is from heaven and which flows after being dug, and even that spring by itself, the bright pure water which leads to the sea, may the divine water protect me here.*

Water is the upholder of all lives and the savior of everything existing on earth. However, this life-giving substance, when contaminated, causes outbreak of disease and destruction of eco-system. In 2002, United Nations Committee on Economic, Social and Cultural Rights stated, 'Water is fundamental for life and health. The human right to water is indispensable for leading a healthy life with dignity. It is a prerequisite to the realization of all other human rights'. Water has very strong influence on human health both as a means to controlling disease and also as a medium through which disease causing agents may be transmitted. The adverse impact of water on health derives principally from the consumption of water containing pathogenic organisms or toxic chemicals and the inadequate intake of water that leads to poor hygiene. *United Nations Environmental Programme Report* states that over half of the world's hospital beds are occupied with people suffering from illness linked with contaminated water and more people die as a result of polluted water than are killed by all forms of violence, including wars. The statistics are stark: globally, everyday two million tons of sewage, industrial and agricultural waste is discharged into the world's waterways and at least 1.8 million children under the age of five years die every year from water related disease, or one every 20 second.

Continuing population growth and the changing production and consumption patterns have resulted in an upsurge of industrial processes, mining, agriculture and urbanization – with the resultant release of heavy metals, organic toxins and discarded pharmaceuticals into the environment. Climate change has further precipitated the problem. This presents a global threat to human health and well-being.

India, a water adequate country, a country of 'Sujalam Sufalam', is also facing several conflicts and fierce competition from different stakeholders. Demand-supply imbalance has led to precarious condition of water resources. Human settlements, disposal of sewage and industrial effluents are causing degradation of water bodies. A whopping 90 per cent of raw sewage and 70 per cent of untreated industrial wastes are being dumped into surface waters. In arid regions, use of canal irrigation has led to the problem of water logging and salinity. In the course of its travel through typical rock structures in the sub-surface, naturally occurring elements like arsenic and fluoride also get dissolved in water.

The present volume compiles studies mainly focusing on the health impacts of contaminations like fluoride, nitrate, and salinity; pollution caused by pesticide and heavy metals; and lack of safe drinking water and sanitation. This book is divided into four sections. The first section, which includes three research papers, examines the impact of pesticides on human health. The papers have described certain characteristics of persistent organic pollutants used in the form of pesticides in agricultural activities and have explained various routes mainly the food chain through which they enter human body and cause harm to health.

The second section is dedicated to 'Fluoride and Health'; a problem which has become endemic in India and particularly in Rajasthan, hence requires immediate attention. Four papers broadly present the status of fluoride at the national level and in the state of Rajasthan; discuss ill-effects of fluoride mainly on teeth and skeleton deformation, endocrine, neurological, reproduction and effect on immune system, etc.; suggest some mitigation measures and also underline a few de-fluoridation methods.

In the third section, the focus turns to 'Nitrate, Heavy Metals and Salinity' – all three despite being major contaminants of ground and surface water have not evoked much research mainly due to lack of data. One of the three papers included in this section, 'Environmental Health Aspects of Nitrate Toxicity in Humans and its Mitigation Measures' discusses the exogenous, endogenous and other sources of nitrate pollution which include agriculture, municipal and industrial source of pollution. It further throws light on different kinds of cancer, problems caused in the cardiovascular system that include early onset of hypertension, inflammation and degradation of cardiac musculature, endothelial dysfunction, atherosclerosis and gastrointestinal system, and a number of other diseases like diabetes, thyroid, adrenal gland, and immunity which are affected by nitrate/nitrite ingestion. Two research papers using case studies discuss about problems and diseases caused by salinity and heavy metals respectively.

Last section is on 'Sanitation and Health', this is an area that has elicited attention recently – both among the policy makers and among scholars. Two papers present broad overview of access to safe water and sanitation in Rajasthan and regional variations therein; linkage of drinking water and sanitation and health using district level data of Rajasthan. Other three papers discuss anthropogenic problems related to ground water, climatic change situation in the Banas Basin using temperature and precipitation data, and ground water pollution scenario in Rajasthan State, respectively.

Hopefully, the book will draw attention of academicians and policy makers to the alarming situation and will inspire firm actions to save human life and ecosystem. It is far cheaper to protect water resources than to clean up polluted water. The ability of ecosystems to provide services has also reduced considerably. To succeed in the face of some of the largest threats to human health and environmental degradation, an integrated approach to freshwater and waste water management is required, which involves innovative and region specific pathways along with community participation. To attain better development outcomes, water must be made an integral part of all planning and management.

Section–I
Pesticides and Health

2

Persistent Pollutants and Human Health

Pradeep Bhatnagar and Sreemoyee Chatterjee

Evidence demonstrates that certain chemicals are used in the production of many items that we use every day. Some of these organic chemicals are persistent, bioaccumulative and toxic, causing long-term harm to the health of human beings and the planet's environment. Throughout the world, people and their environment are exposed on a daily basis to Persistent Organic Pollutants (POPs). POPs and their effects on human health and the environment is an issue of global concern. POPs are characterized by their lipophilicity, persistence and semi-volatility. These organic compounds are resistant to photo-catalytic, biological and chemical degradation. Generally POPs are halogenated and show high solubility in fats and have very low solubility in water. These properties make POPs very potent environmental pollutants. Because of their lipophilic nature they are easily accumulated in the organism (bioaccumulation). Since the cell membrane (plasma membrane) contain 30 per cent dry weight of lipids, therefore, these pollutants easily enter in the cells and being hydrophobic in nature they cannot be excreted easily. The organism keeps accumulating these chemicals on every exposure.

Toxins (POPs) that have a low concentration in the environment can reach dangerous levels inside the cells and tissues through this process of bioaccumulation. They are also semi-volatile, therefore, they become mobile and move long distances in the environment before depositions occur. The effects of these POPs are also magnified in the environment through the food chain. Species (animals) not normally in direct contact with polluted (contaminated) environment or media may become exposed through ingestion of contaminated prey, promoting accumulation or magnification of pollutant into higher trophic levels.

This phenomenon is called biomagnification; which occurs when the toxic burden of a large number of organisms at a lower trophic level is accumulated and concentrated within a predator at a higher trophic level. For example in an aquatic ecosystem contaminated with toxic pollutant like dichloro diphenyl trichloroethane (DDT), concentration of which may be 0.0001 ppm in water, planktons and bacteria absorb this toxic POP molecules from sediments or water, but they fail to excrete that toxicant (DDT), therefore, the concentration of the pollutant increases to 0.001 ppm in them. Small fishes which prey on them having 0.001 ppm of toxic chemical (DDT) may accumulate 0.01 ppm of DDT. Big fish which feed on these contaminated small fish will accumulate very high concentration of pollutant say 1 ppm. Their predators – the top carnivores in the food chain – like game fish, fish-eating birds and humans can accumulate high toxin levels such as 1–10 ppm. The phenomenon of bioaccumulation and biomagnification make lipophilic POPs very potent environmental pollutants, though the water body is contaminated with very low concentration of toxicants, but higher animals including ourselves are exposed to 10 ppm of toxicants (Figure 2.1).

Some of the POPs have become extremely widespread, being found now from Tropics to Antarctica. They often accumulate in food webs and reach toxic concentrations in long-living top predators, such as humans, sharks, swordfish and bears. POPs include 12 organic compounds: Aldrin, Chlordane, DDT, Dieldrin, Dioxins, Endrin, Furans, Heptachlor, Hexachlorobenzene (HCB), Mirex, Polychlorinated biphenyls (PCBs) and Toxaphene.

To summarize, POPs have the following three characteristics:

Persistent: POPs are chemicals that last for a long time in the environment. Some may resist breakdown for years and even decades while others could potentially break down into other toxic substances.

Bioaccumulative: POPs can accumulate in animals and humans, usually in fatty tissues and largely from the food they consume. As these compounds move up the food chain, they concentrate to levels that could be thousands of times higher than acceptable limits.

Figure 2.1

Schematic Diagram Showing Bioaccumulation and Biomagnification of Toxic POPs

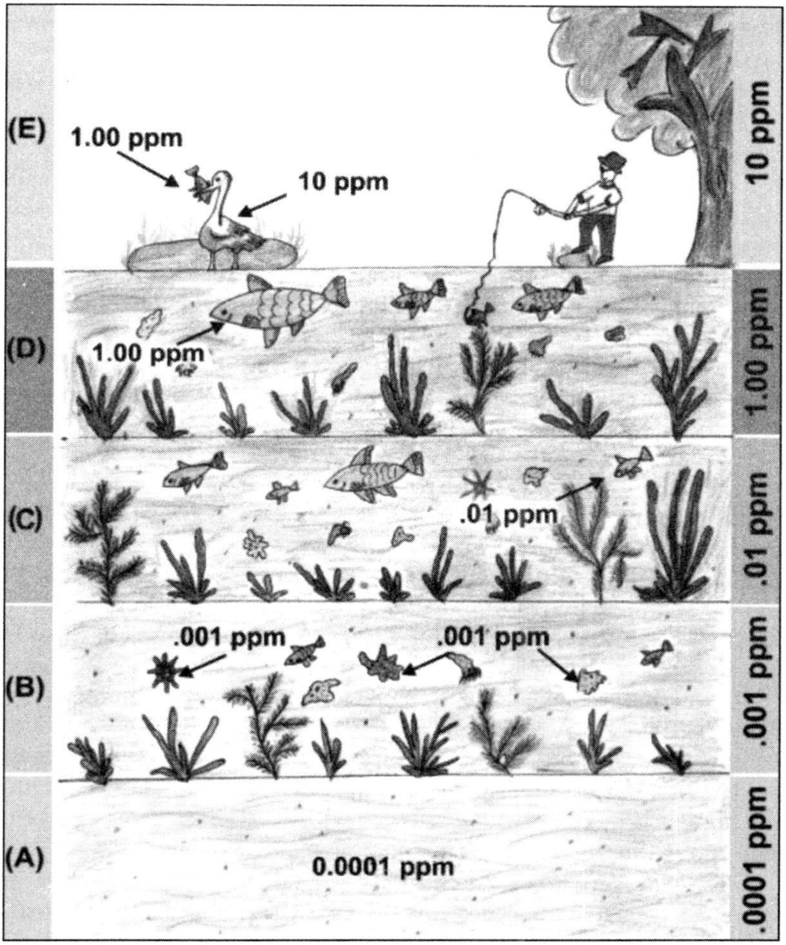

Toxic: POPs can cause a wide range of health effects in humans, wildlife and fish. They have been linked to effects on the nervous system, reproductive and developmental problems, immune response suppression, cancer and endocrine disruption.

Few POPs that are of greatest concern in the current scenario are:

Polybrominated Diphenyl Ethers (PBDE) are widely used as flame retardants in textiles, foams in upholstery and plastics in appliances and computers. It was first reported in Sweden in the 1990s that this compound was accumulating in women's breast milk. It was subsequently found in humans and other species everywhere from Canada to Israel. The toxicity and environmental persistence of PBDE are much like those of PCBs to which it is closely related chemically.

Perfluorooctane Sulfonate (PFOS) and *Perflurorooctanoic Acids (PFOA or C8)* are members of a chemical family used to make non-stick, waterproof and stain resistant products such as teflon, gore-tex and scotch guard. Heating non stick cooking utensils above 260°C may release PFOA, sufficient to kill a small bird. These chemicals which are reported to be infinitely persistent in the environment – are found throughout the world. This chemical family has been shown to cause liver damage, various cancers, reproductive and developmental problems in rats.

Phthalates are found in cosmetics, deodorants and many plastics (PVC) used for food packaging and children's toys. Many phthalates act as endocrine hormone disrupters and have been linked to reproductive abnormalities and decreased fertility.

Bisphenol A (BPA) a prime ingredient in polycarbonate plastics, commonly used for products ranging from water bottles to tooth protecting sealants, has been widely found in humans. So far, there is little direct evidence linking BPA exposure to human health risks, but studies in animals have found that the chemical can cause abnormal chromosome number (aneuploidy) which is the leading cause of miscarriages and several forms of mental retardations. Being an environmental estrogen this may alter sexual development in both males and females.

Atrazine is a widely used herbicide in USA, mainly on corn, cereal grains, golf course, and sugarcane and even on Christmas trees. It is well known as an endocrine disrupter in mammals, resulting in spontaneous abortions, low birthweight and neurological disorders.

Pesticides are toxic chemicals designed to kill pests. They are named on the basis of their target pest organism, e.g. insecticides (insect), fungicides (fungus), nematicide (nematodes), rodenticide (rats), molluscicides (molluscs) and so on. Indiscriminate use of chemical pesticides in various agricultural sectors has emerged as a cause for long-term irreparable environmental damage throughout the world. Many of them contaminate the ecosystem and can be traced in plants and animals. The residual compounds reach the human tissues through eating. Besides causing acute and chronic toxicity these pesticides are affecting the immune system in general and other sensitive systems leading to immune suppression, immune potentiation and hypersensitivity of the host against infectious diseases. Here we are more concerned with insecticides because large amount of chemicals are used in agriculture sector for the protection of various crops, vegetable and horticulture from a variety of insect pests. These insecticides can be categorized into four chemical groups.

Organochlorine (chlorinated hydrocarbon): This class of insecticide also refers to I generation insecticides. These insecticides were discovered during 1939–1945 war. DDT and BHC (Benzene hexachloride or Hexachlorocyclohexane) now known as HCH (hexachlorohexane) were at first praised as almost perfect insecticides – being relatively cheap to manufacture, strikingly effective against numerous insect pests and yet apparently safe to men and other warm blooded animals (birds and mammals). Since that time several billions of people have been chronically exposed to DDT. Yet outside the range of industrial accidents or attempted suicides (acute exposure), chronic exposure did not show any serious medical side-effects during early period (for exposure up to 60 sec.). Conversely, probably no chemical made by man, not even penicillin, streptomycin and sulphonamides has saved so many lives as has DDT. The reason for this is that it was for some time outstandingly successful in controlling the vectors of organisms responsible for such mortal and debilitating diseases as malaria, river blindness and yellow fever.

Other important members of this persistent group of insecticides are: Endosulfan, Aldrin, Dieldrin, Diclofol, Toxaphen, Chlordane, Heptachlor and Methoxychlor. All members of this group are highly lipophilic, therefore, they are persistent in the environment and are generally neurotoxic; affect both peripheral (DDT) and central nervous system (BHC and Aldrin). Another important quality of members of this group is that they show low acute toxicity in non-target organisms (birds and mammals) as they do not cause immediate health effects. We are concerned about chronic toxicity, because even low dose of these insecticides when exposed to over a lifetime, is critical. What we need to learn is how these insecticides would slowly, invisibly damage and hurt the body's immunity or cause other health disorders. Endosulphan (Cyclodiene organochloride) is a broad spectrum organochlorine (OC) insecticide and is widely used in agricultural countries, like India. Widespread health effects on animals at the top of the food chain, caused by organochlorine pesticides and other persistent chemicals have been documented to have reached epidemic and catastrophic levels. Ample evidence that organohalogens have detrimental effects on the immune system of man and animals has been collected over the past three decades. Marine mammals and humans, often at the top of the aquatic food chain, accumulate multiple environmental contaminants such as organohalogens that are now ubiquitous in the aquatic ecosystem. The dysregulation of immune functions by some pesticides leads to various immune disorders, including immunodeficiency, tumorigenesis, allergies and autoimmunity. Genotoxicity is a serious possible harmful effect caused by these compounds and calls for special attention in view of the irreversible nature of its manifestation. Excessive use of OC insecticides causes indiscriminate killing of beneficial as well as harmful insects and insect resistance is developed (Carson, 1963) which led to the development of other groups of insecticides (II and III generation). They do not remain in the environment for a longer period, i.e. they are not persistent like OCs. This important group of insecticides are:

Organophosphates: They are also known as II generation insecticides. Important members of this group are Melathion, Parathion, Monocrotophos Chlorpyrifos, Phosphamidon, Diazinon and Dichlorvos. These are highly neurotoxic. They are of concern for children's health. These pesticides are frequently applied to many foods important in children's diets, and certain Organophosphate pesticide residues can be detected in small quantities on these foods. Organophosphates can interfere with the proper function of the nervous system when exposure is sufficiently high (Eskenazi, 1999). Childhood is a period of increased vulnerability, because many children may have low capacity to detoxify organophosphate pesticides through age 7 years (Huen, 2009). Though, they are not persistent in nature, but recent studies indicate their residues in the farmers of Punjab, who use these insecticides and are constantly exposed to them. Since 1999, Environment Protection Agency (EPA) has imposed restrictions on the use of organophosphate pesticides Azinphos methyl, Chlorpyrifos and Methyl parathion on certain food crops and around homes, largely due to concerns about potential exposures of children (US EPA, 2002, 2006, 2008). But a study carried out by Centre for Science and Environment, India in 2006 explored the high incidence of cancer in certain areas of Punjab and its linkage with these insecticides.

Carbamates: Members of this group are also very toxic to insects as well as to other animals; they are a powerful inhibitor of the nerve cholinesterase, which is responsible for the transmission of nerve impulse from one neuron to another neuron or to muscles, causing hyperactivity of the muscles resulting in fatigue which is followed by death. Some important Carbamate insecticides are Carbonyl (Sevin), Propoxur, Carbofuran, Aldicarb, Triazine and Oxamyl. These insecticides also do not accumulate in the biological system so they are non-persistent.

Natural and Synthetic Pyrethroids: These are new generation insecticides, cause neurotoxicity like OCs. Few members of this group persist in the environment for a shorter period and have lower mammalian toxicity. They are derived from *Chrysanthemum cinerariaefolium*, a plant which is grown commercially as a source of these insecticides in Kenya and Tanzania. Natural pyrethroids have been of little use for crop protection since they are rapidly destroyed by light. They are most effective against household insects, flies and mosquitos.

The synthetic pyrethroids were introduced in the year 1949. Being light stable they were effective against crop pests. Some of the important synthetic insecticides which are being used nowadays are Permethrin, Fenvalerate, Decamethrin and Pyrethrin. The behavioural and physiological manifestations of poisoning of these insecticides resemble those caused by DDT. In particular, an initial period of sensory hyperexcitation leads to loss of coordination, ataxia, prostration and death.

Movement of POPs (insecticides) in the Environment/Insecticide Cycle: When pesticides are applied on crops, some part of it evaporates and is transported by air to long distances; the remaining part seeps in to the ground and may contaminate groundwater or reach streams or rivers (Figure 2.2).

Insecticides in Food Chain: All the POPs including insecticides (OCs) can travel great distances through the environment. When insecticides are applied on crops or vegetables it may contaminate the food chain (Figure 2.2).

We are exposed to them either directly or through consumption of food (vegetable, wheat, dairy products) or contaminated water (Figure 2.2). As previously discussed they accumulate in our fat cells (being lipophilic).

POPs can move to the developing foetus through the placenta and infants may be exposed to it through breast milk. They easily leach from the mother's fat cells to milk (milk fat). During starvation or diseased conditions these insecticides leach out and affect the vital organs like heart, liver, endocrine glands or even the immune system.

Figure 2.2

Possible Routes of Contamination of Pesticides in Food Chain

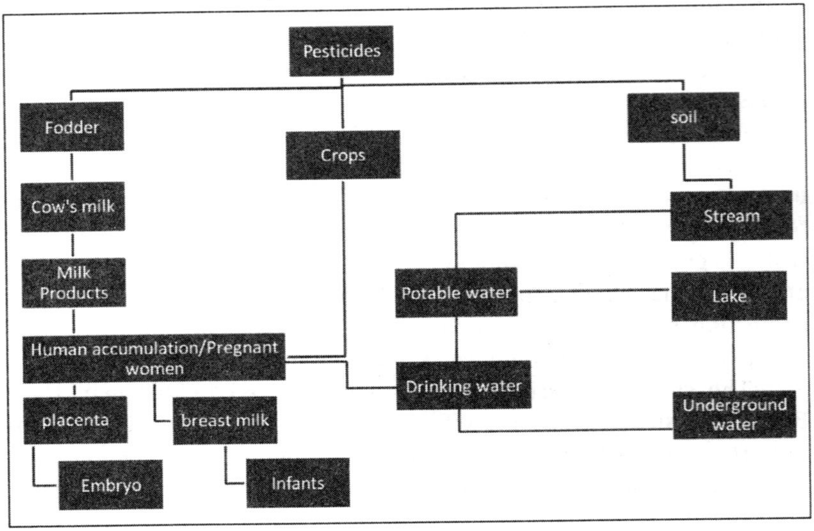

Insecticides in Our Food Items

Insecticides and other POPs have contaminated almost all food items, worldwide. Fruits, vegetables, poultry and milk are all laced with high pesticide residues much above the maximum residue limits (MRL). Samples of Amul milk collected from Ahmedabad, for instance, had the highest traces of Chlorpyriphos, a known carcinogen that can also cause neural disorders. Zee News in February this year exposed an alarming level of pesticides in 509 out of 16,790 samples – comprising vegetables, fruits, spices, rice, wheat and other food items – on the basis of a study conducted in 2013–14 and documented in a government report. The study was conducted as part of the central scheme 'Monitoring of Pesticide Residues at National Level' operational since 2005. The samples were collected from various wholesale and retail markets located at different parts of the country and analysed by 25 laboratories.

We will discuss the level of contamination and its consequences with reference to the city of Jaipur, in the state of Rajasthan.

Vegetables

We all know that green vegetables are good for health. How many times did our parents tell us this? But nowadays fresh, leafy and other vegetables are losing their place on the table. Many studies showed that at several places in India, vegetables are contaminated with various types of POPs including insecticides. In a study conducted by Bakore et al., 2002, samples of vegetables viz. okra, spinach, cabbage, cauliflower, potato and tomato were collected from different markets of Jaipur, at the beginning, in the middle and at the end of the seasons with respect to the particular vegetable. Samples were analysed for organochlorine insecticide residues. Results showed that almost 75 per cent samples were contaminated with pesticides like Aldrin, Parapara dichlorodiphenyl trichloro ethane (p,p'DDT), p,p'-Dichlorodiphenyl trichloro ethylene (p,p'DDE), p,p'Dichlorodiphenyl dichloro ethane (p,p'DDD), isomers of Hexachloro cyclohexane (α, β and γ HCH), Heptachlor and Heptachlor epoxide.

Comparative concentration of total organochlorine insecticides' residues detected from different stages of seasons is depicted in Figure 2.3. Residue levels in majority of vegetable samples were above the maximum acceptable daily intake (ADI) or tolerance limit prescribed by World Health Organization (WHO, 1973). A number of samples have been found to contain insecticide residue above legal limits fixed for vegetables under the Prevention of Food and Adulteration Act, 1954 by Ministry of Health, Govt. of India. The Maximum Residue Limit (MRL) under the PFA Act, 1954 for DDT- 3.5 ppm, HCH (Lindane)- 3.00 ppm, Aldrin- 0.1 ppm and Heptachlor- 0.005 ppm.

Figure 2.3

Total OCP Residues in Different Vegetables during Different Stages of Season

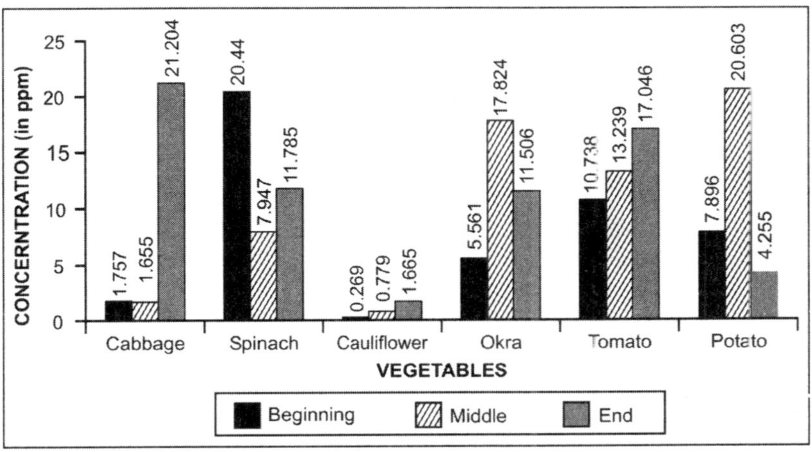

In Jaipur, most of the vegetables sold in the market are grown around Amanisha Nalla (drainage) and are sprayed with a variety of insecticides and washed with the same insecticide contaminated water. Besides insecticides, drainage water may also contain other harmful chemicals. For the first time this study showed that vegetables marketed in Jaipur city are highly contaminated with persistent insecticides.

Milk Supplied to Jaipur

In studies (John and Bhatnagar, 2001; Saxena et al., 2005) samples of dairy milk (toned and whole) and buffalo milk, supplied by vendors during summer, rainy and winter seasons, in Jaipur city were monitored for insecticide residues. The milk samples in all the seasons were found contaminated with insecticides; pp'DDT, pp' DDD, α, β and γ HCH, Heptachlor and Heptachlor Epoxide (Figure 2.4). The level of insecticide residue in the milk was quite high and many of these compounds are reported to be carcinogenic, mutagenic and teratogenic. It was suggested that Pollution Control Board or Agriculture Departments of Rajasthan should monitor the residues of these chemicals on a regular basis.

Figure 2.4

OCP Residues in Buffalo Milk during Different Seasons

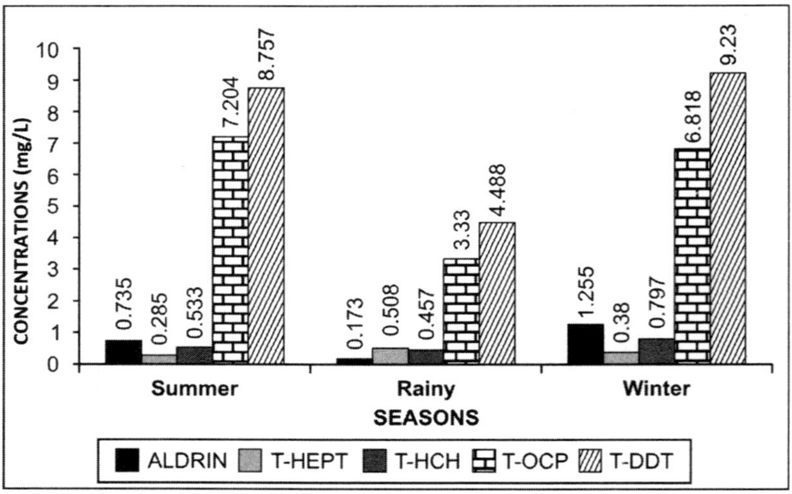

Waterbodies of Jaipur: In order to have baseline information on the levels of some of the commonly used organochlorine pesticide residue, a survey of two lakes, around Jaipur was carried out (Kumar et al., 2006). These lakes were, at that time, characterized by high fish productivity and were important recreational areas, being temporary abodes for many migratory birds. One of the lakes, Jalmahal, still attracts many local migratory birds and has been developed by the Rajasthan Tourism Development Corporation (RTDC) as an important tourist destination.

The survey indicated that both the lakes were contaminated with moderate to high levels of organochlorine insecticides. The average insecticide residues in Mahlon Lake varied from 0.93 to 5.71μgL^{-1} whereas in Jalmahal Lake the range was 0.55 to 11.4 μgL^{-1}. Among the major organochlorine insecticides/metabolites, pp'-DDT, pp'-DDE and pp'-DDD constituted major components in the lakes. These concentrations of organochlorine insecticides are much higher than their recommended limits (2μgL^{-1}) in water for protection of aquatic life (Water Quality Criteria, 1972). The study was perhaps the first one which added to the basic information on the insecticidal pollution of aquatic ecosystem of Rajasthan.

Major carps like *Labeo bata, Wallago attu, Channa punctatus* and *Puntius sarana* of water reservoir had significant amount of organochlorine insecticide residues in their vital organs viz. kidney, muscle, brain, gills and alimentary canal (Bakre et al., 1990). Fish from this lake are generally consumed in Jaipur city.

Human Exposure

From our studies it was established that organochlorine insecticides has entered the food chain of Jaipur area. Human population are getting exposed to these persistent pollutants through vegetables, milk and water or may be through other sources also. Many studies have been carried out on the exposure of the human population of Jaipur city to organochlorine insecticides and their consequences.

Movement of Insecticides from Mother to Foetus and to Infants

A study was carried out in Jaipur city in which insecticide residue levels in the circulating blood, cord blood and in placenta of pregnant women undergoing full term normal delivery, were analysed. All the above samples had significant residue of DDT and its isomers, HCH and its isomers like Heptachlor, Aldrine and Heptachlor epoxide. Total amount of organochlorine insecticides ranged from 10.00 to 9,698.0 ppb in the blood of women, 15.0 to 6,317.0 ppb in the cord blood and 25.0 to 5,734.0 ppb in the placental tissue. Non-significant amount of above insecticide residue was detected in the breast milk of women (Sharma and Bhatnagar, 1996).

Anupgarh region of Rajasthan is known for high agricultural production as well as usage of pesticides. In one of the other study, the blood and milk of lactating women of Anupgarh (North-West Rajasthan) was analysed for insecticide residues. Both blood and breast milk was contaminated with OC insecticide.

Burden of persistent organochlorine insecticides in women of Jaipur area and its relation to hypothyroidism, breast cancer, cervical cancer and blood cancer is well established. In an extensive study, blood samples of 123 women from Jaipur city who visited the thyroid centre of S.M.S. Medical College and Hospital, were analysed for insecticide residue. Twenty-three out of 123 women had depleted T_4 and high TSH levels. Remaining women had normal thyroid hormone levels. Though the level of total organochlorine insecticides was higher in all subjects but level of Dieldrin was found to be much higher in women suffering from hypothyroidism. Total insecticide residue level in the blood of women having normal thyroid hormone was 14.12 ppm while in women having depleted T_4 and high TSH levels, 17.93 ppm of total insecticide was found in their blood (Rathore et al., 2002).

In another study, Jaipur women were assessed for the burden of organochlorine insecticides (DDT and its metabolites DDD and DDE, Dieldrin, Heptachlor and HCH and its isomers α, β and γ). Findings of the study indicated that the organochlorine insecticide residues in breast cancer patients were found to be significantly higher (7.468 mg/L) than normal females (1.857 mg/l) irrespective of their age, diet or geographical distribution.

Contamination of blood with these noxious chemicals is a matter of great concern as low (chronic) dose exposure of these chemicals in human beings may lead to health effects such as hormone disruption, reproductive abnormalities and cancer. Although their chemical structure do not closely resemble estrogen, chlorinated POPs including insecticides strongly mimic estrogen in the body and are characterized as Endocrine Disrupting Compounds (EDCs). Estrogen provides stimulus for genetically altered breast cells to proliferate (divide) more rapidly than normal cells, a process that eventually causes a tumor. Therefore, there are possibilities that higher concentration of POPs in the blood of women may enhance the chances of developing breast and other cancers. In light of increasing rates of breast and other cancers and our limited knowledge of the possible causes, investigation of the carcinogenic environmental chemicals particularly POPs is a promising avenue for further elucidation of the origin of carcinogenicity.

In a study (Mathur et al., 2008), blood samples of 150 women, out of which 100 ladies were suffering from reproductive tract cancers like cervical, uterine, vaginal and ovarian cancers, were collected. The rest 50 that did not suffer from any type of cancer or any major disease were treated as the control group. Blood samples from all the subjects were found laced with various OC insecticides and their isomers, like HCH (BHC), DDT, Aldrin and Heptachlor. In the patients with reproductive tract cancers the total burden of OC insecticide was 8.71 mg/l. The total insecticide burden in cervical cancer patients was 10.891 mg/l, in uterine cancer patients was 7.506 mg/l and 6.303 mg/l was observed in vaginal cancer patients followed by ovarian cancer patients with a total pesticide burden of 7.577 mg/l. Levels of insecticide in cancer patients were found to be significantly higher than the control subjects. Studied cases were not exposed to the insecticides directly, which could imply that the subjects were exposed to these POPs through food containing insecticide residue.

In a study (Gupta, 2007), 285 blood samples from human subjects residing in Jaipur region, were analysed for 15 OC insecticides (HCH and its isomers, DDT and its isomers, Aldrin, Heptachlor, Heptachlone Epoxide, Endosulphan I and II and Endosulfan sulphate). Around 150 subjects suffered from leukaemia, 55 subjects suffered from lymphoma and 80 subjects did not suffer from any cancer (control group). In this study the association between the risk of leukaemia and lymphoma and residue level of insecticide was examined. A significant high concentration of total OC insecticide in the blood of leukaemia patients (0.4772 mg/l) and lymphoma patients (0.4874 mg/l) was observed in comparison to healthy person (0.3456 mg/l). Slightly higher levels of insecticide residue in the blood of cancer patients indicated a positive relationship between disease and Organochlorine pesticides (OCPs). Occurrence of OC insecticide residue in blood of general population confirms the exposure of people to the insecticides either directly through occupational practices, application in fields and houses, or indirectly, by consuming contaminated water or food.

Environment contaminated with POPs like OC insecticide's residues may be an important etiological factor in reproductive tract or other cancers like leukaemia and lymphoma. Given the widespread dissemination of OC insecticides in environment and in the food chain, the implications are far-reaching for public health intervention worldwide. Stricter regulation may be deployed and such measures have to be weighed against the benefits of insecticide use.

In previous pages we discussed the studies with reference to OC insecticides, monitoring of food commodities and human population of Jaipur region only. We have also tried to find out their relation with the occurring endocrine disorders and cancers in overall human population.

At the national and international level as well, insecticides have been extensively monitored and studied for their relation with occurrence of many problem related to health. Still very little is known about the direct link between POPs body burden and the consequent health impact. Bio monitoring these chemicals in blood, urine, breast milk, fat or in other tissues, do reveal that large amount of POPs residues have infiltrated our bodies. Yet some scientist and industry persons argue about their links with human health. According to the industries, there is a lack of 'authentic' epidemiological studies correlating diseases to one particular chemical, since many different types of pesticides and other POPs chemicals have invaded the human body.

But the toxicological impact of individual POPs, particularly insecticides is stark. Till date there is no evidence that proves the safety of carrying chemical body burden; in fact there is growing evidence to incriminate such a burden. Bio monitoring research suggests that the chemical cocktail in the human body could be linked to a host of afflictions like developmental disorders, fertility problems, neurological problems and learning disabilities in children, immunological problems and cancers. In fact, we do not know about cumulative health impact of dozens of chemicals present in the body at the same time.

Alternatives

Some alternatives for reducing our dependence on dangerous chemicals particularly insecticides, include management changes, such as using cover crops and mechanical cultivation, planting mixed poly-culture rather than vast monoculture fields and organic conventional farming. We should learn to accept less than perfect vegetables and fruits, if we want to avoid toxic chemicals in our diet, because many synthetic chemicals are applied on vegetables and fruits to enhance their cosmetic looks.

Biological controls, such as insect predators, pathogens, or natural poisons specific for a particular pest, can help to reduce chemical use. Genetic breeding and biotechnology can produce pest resistant crop and livestock strains as well. Integrated pest management (IPM) combines all of these alternative methods, together with judicious use of synthetic pesticides under precisely controlled conditions.

Reducing the Insecticides in Our Food

We should follow some simple rules for reducing the amount of pesticide residues and other toxic chemicals in our diet:

- Washing and scrubbing all fresh fruits and vegetables thoroughly under running water.
- Peeling fruits and vegetables when possible. Outer leaves of leafy vegetables, such as cabbage and lettuce should be discarded.
- Cooking foods to break down chemical residues.
- Trimming the fat from meat, chicken and fish. We should try to eat lower on the food chain, when possible, to reduce bio accumulated chemicals.
- Growing our own fruits/vegetables without using insecticides or with minimal use of dangerous chemicals.
- Asking for organically grown food at our local grocery stores.

References

Bakore, N., John, P.J. and Bhatnagar, P. (2002). 'Evaluation of Organochlorine Insecticide Residue Levels in Locally Marketed Vegetables of Jaipur City, Rajasthan, India'. *Environment Biology*. 23(3): 247–252.

Bakre, P.P., Misra, V. and Bhatnagar, P. (1990). 'Residues of Organochlorine Insecticides in Fish from Mohala Water Reservoir', Jaipur, India. *Bulletin of Environment Contamination and Toxicology*. 45: 394–398.

Carson, R. (1963). *Silent Spring*, Hamish Hamilton, London.

John, P.J., Bakore, N. and Bhatnagar, P. (2001). 'Assessment of Organochlorine Pesticide Residue Levels in Dairy Milk and Buffalo Milk from Jaipur City', Rajasthan, Jaipur, *Environment International*. 26: 231–236.

Kumar, A., Baroth, A., Soni, I., John, P. J. and Bhatnagar, P. (2006). 'Organochlorine Pesticides Residues in Milk and Blood of Women from Anupgarh', Rajasthan, India. *Environment Monitoring and Assessment*. 116: 1–7.

Kumar, S., Lal, R. and Bhatnagar, P. (1988). 'Residues of Organochlorine Insecticides in Two Lakes of Jaipur', India. *Water Air Soil Pollution*. 42: 57–65.

Mathur, V., Bhatnagar, P., Sharma, R. G., Acharya, V. and Saxena, R. (2002). 'Breast Cancer in Relation to Pesticide Exposure in Women of Jaipur'. *Environment International*. 28: 331–336.

Mathur, V., John, P.J., Soni, I. and Bhatnagar, P. (2008). 'Blood Levels or Organochlorine Pesticide Residues and Risk of Reproductive Tract Cancer among Women from Jaipur', India. *Advances in Experimental Medicine and Biology*. Vol. 617.

Rathore, M., Bhatnagar, P., Mathur, D. and Saxena, G.N. (2002). 'Burden of Organochlorine Pesticides in Blood and its Effect on Thyroid Hormone in Women'. *Science Total Environment*. 295: 207–215.

Saxena, G., Agarwal, M., Baroth, A. and Bhatnagar, P. (2005). 'Pesticide Residues in the Milk of Jaipur City'. *Ecophysiology Occupational Health*. 5: 77–79.

Sharma, M. and Bhatnagar, P. (1995). 'Effect of Persistent Pesticide Pollution on the Health of Women and Neonates', *Health Hazards, Gender and Society*, Indu Mathur and Sanjay Saxena (eds). Rawat Publications, Jaipur.

Sharma, M. and Bhatnagar, P. (1996). 'Organochlorine Pesticides and Preterm Labour in Human Beings'. *Current Sciences*. 71(25): 628–631.

US Environmental Protection Agency. (2002). Child Specific Exposure Factor Handbook 2002 (Interim Report). http://cfpub-epa.gov/ncea/cfm/recordisplay.cfm?deid=55145.38.U.S.

——. (2006). Child Specific Exposure Factor Handbook 2006 (External Review Draft). http://cfpub-epa.gov/ncea/ cfm/recordisplay.cfm?deid=56747.

——. (2008). Child Specific Exposure Factor Handbook 2008 (Final Report). US Environmental Protection Agency, Washington, DC, EPA/600/R-06/096F, 2008.

3

Detrimental Outcome on the Human Health with the Emergent Usage of Pesticides

Ashu Chaudhary, Anshul Singh,
Priyanka Sharma and S.C. Joshi

Pesticides are substances or a mixture of substances, of chemical or biological origin, used by human society to mitigate or repel pests such as bacteria, nematodes, insects, mites, molluscs, birds, rodents and other organisms that affect food production or human health. They usually act by disrupting some component of the pest's life processes to kill or inactivate it. The term pesticide covers a wide range of compounds including insecticides, fungicides, herbicides, rodenticides, molluscicides, nematicides, plant growth regulators and others (Aktar, 2008). A fundamental contributor to the Green Revolution has been the development and application of pesticides for the control of a wide variety of insectivorous and herbaceous pests that would otherwise diminish the quantity and quality of food produce. Despite their benefits, pesticides can be hazardous to both humans and the environment. Countless chemicals are environmentally stable, prone to bioaccumulation, and toxic (Fenik et al., 2011). Because some pesticides can persist in the environment, they can remain there for years. Environmental contamination or occupational use can expose the general population to pesticide residue, including physical and biological degradation of the products present in the air, water and food (Mostafalou and Abdollahi, 2013).

Figure 1

Detrimental effects of Pesticides

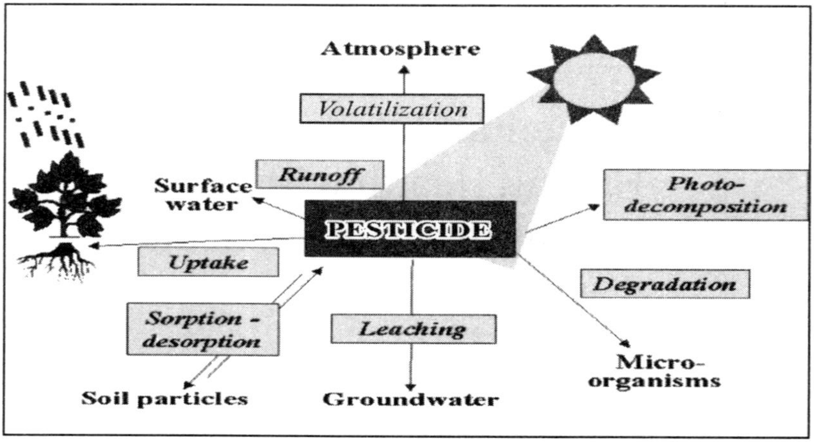

Less than 1 per cent of the total amount of pesticides applied for weed and pest control reaches the target pests. A large quantity of pesticides is lost via spray drift, off-target deposition, runoff and photo degradation, for instance, which can have undesirable effects on some species, communities, or ecosystems as a whole, as well as on humans (Hernandez et al., 2013). Another relevant factor is that low concentrations of many chemicals may not elicit acute detectable effects in organisms, but they may induce other damages, like genetic disorders and physiological alterations, which reduce life span in the long run (Poletta et al., 2009). Ideally, a pesticide must be lethal to the targeted pests, but not to non-target species, including man. Unfortunately, this is not, so the controversy of use and abuse of pesticides has surfaced. The rampant use of these chemicals, under the adage, 'if little is good, a lot more will be better' has played havoc with human and other life forms.

Classification of Pesticides

There are various ways to group pesticides; including classification based on the pests they control (Matthews et al., 2008). The most common classes of pesticides include:

- *Fungicides:* pesticides that kill fungi (including blights, mildews, molds, and rusts).
- *Herbicides:* chemicals that severely interrupt a plant's normal growth process.

- *Insecticides:* pesticides designed to kill insects and other arthropods.
- *Algicides:* pesticides used to control algae in lakes, swimming pools and water tanks.
- *Antifouling agents:* chemicals that repel organisms that attach to underwater surfaces.
- *Antimicrobials and biocides:* chemicals that kill microorganisms such as bacteria, fungi and viruses.
- *Disinfectants and sanitizers:* chemicals used to disable disease-producing micro-organisms on inanimate objects.
- *Fumigants:* a gas or vapour intended to destroy pests in hard to reach areas such as the interior of buildings or soil.
- *Miticides:* pesticides designed to kill mites that feed on plants and animals.
- *Molluscicides:* pesticides used to kill or repel snails and slugs.
- *Nematicides:* pesticides that kill microscopic, worm-like organisms that feed on plant roots (nematodes).
- *Ovicides:* chemicals used to kill eggs of insects and mites.
- *Piscicides:* chemicals for the control of fish.
- *Repellents:* chemicals that repel pests such as mosquitoes or animals.
- *Rodenticides:* pesticides used to control mice and other rodents.

Pesticides grouping can also rely on their chemical structure (EPA 2004):

- Organophosphorus (chlorpyrifos and diazinon)
- Carbamates (carbaryl and aldicarb)
- Organochlorine (DDT and aldrin)
- Pyrethrins
- Pyrethroids (cyfluthrin and cypermethrin)
- Benzoic acids (dicamba)
- Triazines (atrazine and simazine)
- Phenoxyacetic derivatives
- Dipyridyl derivatives (diquat and paraquat)
- Glycine derivatives (glyphosate)
- Dithiocarbamates (maneb and ziram).

Pesticides that bear similar chemical structures exhibit similar mechanism of toxicity and physicochemical properties, as well as comparable transport properties.

Production and Usage of Pesticides in India

Agriculture is the lynchpin of the Indian economy and contributes 18 per cent to the GDP. Ensuring food security for more than 1.27 billion

Indian population with diminishing cultivable land resource is a herculean task. In the process of achieving the target, pesticides play an important role in Indian agriculture. Pesticides, the agrochemicals, are one of the invaluable inputs in sustaining the agricultural production as substantial food production is lost due to insect pests, plant pathogens, weeds etc. (Bhardwaj and Sharma, 2013). The production of pesticides started in India in 1952 with the establishment of a plant for the production of BHC near Calcutta (now Kolkata) and India is now the second largest manufacturer of pesticides in Asia after China and ranks twelfth globally. There has been a steady growth in the production of technical grade pesticides in India, from 5,000 metric tonnes in 1958 to 1,02,240 metric tonnes in 1998. In 1996–97 the demand for pesticides in terms of value was estimated to be around ₹ 22 billion (USD 0.5 billion), which is about 2 per cent of the total world market. The pattern of pesticide usage in India is different from that for the world in general. As can be seen from Figure 3.1, in India 76 per cent of the pesticide used is insecticide, as against 44 per cent globally (Mathur, 1999). The use of herbicides and fungicides is correspondingly less heavy. The main use of pesticides in India is for cotton crops (45%), followed by paddy and wheat.

Figure 3.1

Consumption Pattern of Pesticides

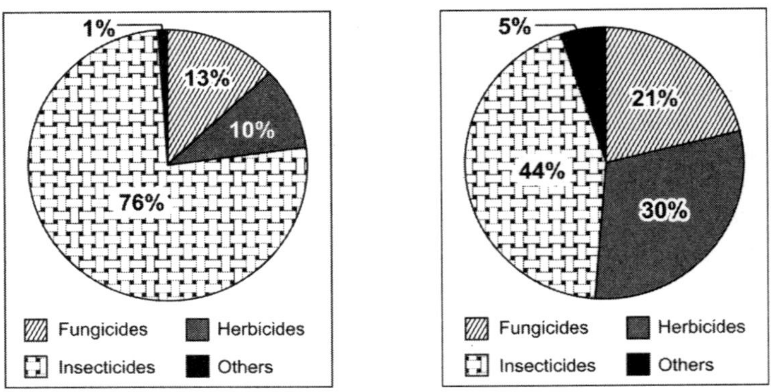

(a) India *(b) World*

Pesticides and Human Health

Presently pesticides are present as common contaminants in the biosphere and on non-target organisms in our urban landscapes, where they can affect plants and animals ranging from beneficial soil micro-organisms and insects, non-target plants, fish, birds and other wildlife. There is now overwhelming evidence that some of these chemicals do pose a potential risk to humans and other life forms and have unwanted side-effects on the environment. One of the consequences of indiscriminate use of pesticide is the adverse health impact on society in general and vulnerable population, like children, in particular. Among human beings some of the well-known health effects of pesticide exposure include acute poisoning, cancer, neurological effects and reproductive and developmental harm.

A pesticide's toxicity may vary between different kinds of organisms and can be affected by a wide range of factors, including how the pesticide enters the body. Exposures occur through the skin, by inhalation, by swallowing, or by eye contact. The physical and chemical differences in product formulations differ greatly in the exposure risk they present for the same active ingredient. Exposure is also affected by whether or not a person is working with the concentrated undiluted pesticide or with the diluted spray mixture during application.

Routes of Entry into the Body

Pesticides enter the human body through the skin, the eyes, the lungs, and/or the mouth. Understanding these four routes of exposure will help pesticide applicators protect themselves from pesticide hazards (Figure 3.2).

Skin or Dermal Exposure

Most pesticide exposures take place through the skin. One study found that approximately 97 per cent of pesticide exposure encountered during spraying operations is skin exposure. The skin is a durable barrier that protects other body organs and systems. This barrier can be injured by toxic chemicals that burn, cause rashes, or enter through scratches or cuts. Skin is also porous, and water, nutrients and some toxic substances can pass through it, affecting organs and systems throughout the body.

Dermal exposure can occur through airborne dust, splashes, spills or spray mist during mixing, loading or applying pesticides. Skin exposure also results from touching pesticide residues or treated surfaces and contaminated clothing, pesticide containers or equipment used during application. Methods such as overhead application, blower application of mists or dusts and animal pour-on or dips are more likely to expose the skin to pesticides.

The pesticide's formulation affects how much the skin can absorb. Dusts, granules, powders and water-soluble liquids and powders do not penetrate the skin as easily as oil-based liquids or concentrated products meant to be mixed with liquids. Not all skin absorbs pesticides in the same way. Warm, moist skin (in the groin, armpits, head, neck or ears) tends to absorb more pesticides than the arms, legs or hands. However, arms and hands are more likely to be cut or scratched, allowing pesticides to move quickly through the break in the skin.

Eye Exposure

The human eye contains many blood vessels close to the surface and can quickly absorb significant amount of pesticide. Eye exposure arises from airborne pesticides (droplets, aerosols or particles) and rubbing the eyes with contaminated hands or clothes. The exposure can damage the eye on contact, especially if the pesticide is corrosive, causing serious burns. Pesticides absorbed through the eye enter the bloodstream and move to other organ systems.

Breathing or Inhalation Exposure

It is easy to breathe in powders, fumes or aerosols (very small droplets suspended in the air) while handling pesticides, especially if people are mixing or loading in a confined space (inside a closed vehicle, a shed or a garage). Breathing in pesticides may also happen outdoors, such as when applicators stand downwind when mixing or loading pesticides into equipment or during application from drift.

Exposure by inhalation can occur in three major ways: the pesticide comes into contact with the nose, throat or lung tissue and causes damage; inhaled pesticide moves into the bloodstream from the lungs and affects other organ systems; or when inducing vomiting to remove a petroleum-based emulsifable concentrate or other pesticide from the stomach, the vomit is accidentally pulled into the lungs, causing lung damage and choking.

Mouth or Oral Exposure

Pesticide that enters the mouth (or is swallowed) is absorbed in the intestinal tract, moving rapidly into other organ systems. Exposure by mouth is most common among children, who put their fingers, objects, or food (including contents of pellet-type bait traps) into their mouth. Adults can become exposed when pesticides splash into the mouth or on lips; when food beverages or cigarettes are consumed after applying a pesticide, or when leftover pesticides are stored in containers with food or beverage labels and are consumed by accident. Other oral exposures occur when people try to suction liquid to clear clogged tubing or nozzles while transferring liquid from one container to another.

Figure 3.2

Routes of Entry of Pesticides into the Human Body

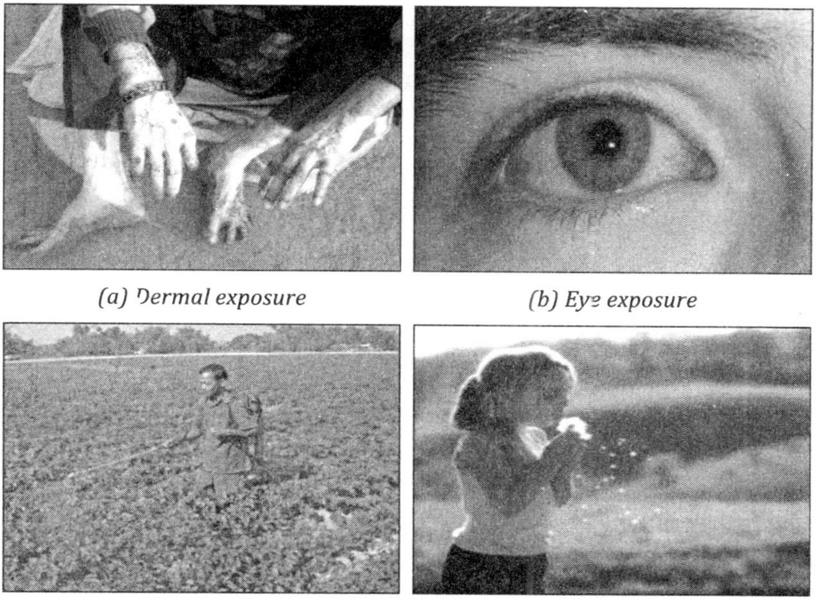

(a) Dermal exposure *(b) Eye exposure* *(c) Inhalation of pesticides* *(d) Oral exposure*

Harmful Effects of Pesticides

Pesticide applicators can experience any one or a combination of these exposures, which may lead to either acute (short-term) effects or chronic (long-term) health effects. Key factors in determining the type and degree of harm include pesticide toxicity, dose, length or frequency of exposure and route of entry into the body.

Acute Health Impact

Acute toxicity refers to the immediate effects of a particular dose of pesticide on human health (Zweiner and Ginsburg, 1988). The symptoms of acute pesticide poisoning depend upon how toxic the chemical is, the route of exposure and the dose that enters the body.

These symptoms can be *contact effects* (affecting skin and external tissue at the entry route), *systemic effects* (affecting internal body systems remote from the entry route) or *allergic effects* (affecting the immune system).

Contact Effects

Injury at the point of contact includes skin discolouration and irritation (dermatitis) such as itching, redness, rashes, blisters and burns. Contact effects may also include swelling, stinging and burning of the eyes, nose, mouth or throat. Many herbicides and fungicides cause dermatitis. Fumigants can cause severe blisters. As herbicides are most commonly applied pesticide group and they predominantly cause contact injury, therefore contact skin effects are the most common form of pesticide injury or poisoning to applicators. Herbicides, fungicides, insecticides and fumigants may cause severe eye irritation or injury, sometimes resulting in irreversible damage. Swelling, stinging and burning of the eyes, nose, mouth or throat are relatively common contact symptoms. Permanent respiratory damage occurs less often.

Systemic Effects

Injury or harm that occurs at sites other than the entry point into the body is called systemic effects. Systemic poisoning symptoms among humans occur most often during application of pesticides that target animals. Common physical symptoms of systemic injury include:

- Nausea, vomiting, diarrhea or stomach cramps (gastric system)
- Headache, dizziness, weakness or confusion (nervous system)
- Excessive sweating, tearing of eyes, chills or thirst (hormonal system)
- Chest pains (cardiovascular system)
- Breathing difficulties (respiratory system)
- Body aches and muscle cramps (musculoskeletal system)

Allergic Effects

Allergic effects are harmful effects some people develop in reaction to substances that do not cause the same reaction in most other people. Acute allergic reactions to pesticides can be systemic, contact or both. The differences are that these reactions happen to a limited number of people with hypersensitivity: immune systems that react strongly to the specific chemical. Pesticide allergy symptoms may be similar to other allergy symptoms – reddening of the eyes, itchy eyes, respiratory discomfort and asthma like symptoms. An allergic reaction may also cause dermatitis, blisters, hives and itching of the eyes. It could also cause illness, asthma-like symptoms or life threatening shock.

As most of these symptoms are similar or identical to those caused by other illness, acute pesticide poisoning is often misdiagnosed.

Chronic Health Impacts

A chronic effect develops over a long period of time, and may last for several years after initial exposure. The effect may be related to long-term or repeated exposure to a pesticide at a low dosage, or to exposure to a high dosage for a short time (Pernille et al., 1998). Types of chronic health effects from pesticide exposure of concern include:

- Cancer or other tumours
- Infertility
- Birth defects
- Nervous system conditions, hormonal imbalances and blocking of natural responses of the immune system.

Pesticides and Cancer

Numerous pesticides are implicated in causing or promoting many types of cancers, leukemia, and lymphomas. Some of these diseases are relatively common, others quite rare. Many of the neoplasms for which association with pesticides is most well-established are among those cancers increasing in incidence in industrialized countries.

The mechanisms by which pesticides contribute to cancer causation vary and one pesticide may operate by more than one of the major mechanisms, which include:

- *Genotoxic effects:* producing direct changes in DNA.
- *Promotion:* causing fixation and proliferation of abnormal clones. This process includes endocrine effects that may stimulate otherwise quiescent but hormonally sensitive cells to carcinogenesis.
- *Immunotoxic effects:* disturbing the body's normal cancer surveillance mechanisms.

Exposure to the pesticides may lead to cancer and various types of cancer which develops in humans are as follows:

Non-Hodgkins Lymphoma (NHL)

Sometimes called the 'silent epidemic,' over the last several decades Non-Hodgkins Lymphoma (NHL) incidence has been increasing by 3–4 per cent per year throughout most of the world (Devesa and Fears,

1992 and Hartge et al., 1994).). In some studies annual increases in incidence are as high as 4.2–8.0 per cent (Cartwright, 1999; Pollan et al., 1998 and Hjalgrim et al., 1996). These reported increases are corrected for known viral causes of NHL, such as human immunodeficiency virus (HIV), and therefore, largely exclude AIDS-related lymphomas (Hartge and Devesa, 1992).

Although the phenoxy herbicides and their contaminants are the most consistently NHL-associated chemicals, investigators raise concern about other pesticides, including lindane (used in some head and body lice-treatments) (Blair et al., 1998), organophosphate pesticides (Zahm et al., 1993) and a variety of others, such as carbaryl, chlordane, dichloro diphenyl trichlorethane (DDT), diazinon, dichlorvos, malathion, nicotine, and toxaphene (Cantor et al., 1992). Evidence shows that some fungicides may also be lymphomagens (Hardell and Eriksson, 1999).

Multiple Myeloma (MM)

Multiple Myeloma (MM) is another hematological malignancy for which age-adjusted incidence seems to have increased during the last several decades. Rates vary, even among industrialized countries: US investigators found an increased incidence of 4 per cent per year from the late 1940s to the early 1980s among white men and women (Devesa et al., 1987). In contrast, epidemiologists in Spain observed a greater than 10 per cent annual rise from the 1960s to the mid-1980s (Pollan et al., 1993). A number of reports cite intermediate increases in several other nations (Cuzick, 1994). Many epidemiological studies reveal an association between employment in farming and the chance of contracting MM, with risks as high as 5-fold (Pottern et al., 1992; Demers et al., 1993; Viel and Richardson, 1993; Kristensen et al., 1996 and Khuder and Mutgi, 1997). Some investigators have more specifically identified possible causative agents. One study of herbicide applicators reports an eight-fold increase in risk of succumbing to MM (Swaen et al., 1992). The phenoxy herbicides are implicated in this excess risk (Eriksson and Karlsson, 1992.), an association that should not be surprising since the malignancy is closely related to lymphoma. Chlorinated insecticides are also associated with increased risk for MM (Nanni et al., 1998).

Myelodysplastic Syndrome (MDS)

Myeloid leukemia and Myelodysplastic Syndrome (MDS) have been associated with occupational exposure to pesticides (Pasqualetti et al.,

1991 and Ciccone et al., 1993). It has been found that there is significant association between occupational exposure to pesticides and both acute myeloid and lymphoid leukemia (Richardson et al., 1992). One cohort study of a group of gardeners known to have been highly exposed to pesticides reveals a nearly 3-fold increased risk for chronic lymphocytic leukemia (Hansen et al., 1992), an illness for which few possible causes have been proposed.

Soft Tissue Sarcomas (STS)

Occupational exposure to phenoxy herbicides and chlorophenol is repeatedly linked to STS (Hoppin et al., 1998). In one of the most detailed investigations of any tumor/pesticide association, one case control study of workers with STS derived odds ratios for exposure to three major pesticide classes- phenoxy herbicides, chlorophenols, and dioxins. The odds of contracting STS after exposure to any phenoxy herbicide was approximately ten times higher than for non-exposed controls; to the class comprising 2, 4-dichlorophenoxyacetic acid, 2, 4, 5-trichlorophenoxyaceticacid, and 4-chloro–2-methylphenoxyacetic acid and to any chlorinated dibenzodioxin or furan, nearly six; and to 2, 3, 7, 8-tetrachlorodibenzo-p-dioxin, greater than five (Kogevinas et al., 1995).

Skin Cancer and Cancer of the Lip

While elevated risk for skin cancer and cancer of the lip is repeatedly associated with farming (Schreinemachers et al., 1999), ultraviolet light exposure may be a more likely causative factor than pesticides. Therefore, observation of an association between one specific type of skin cancer – Bowen's disease – and the manufacture of paraquat (Hsieh et al., 1994) is of interest because the paraquat-associated skin cancers demonstrate DNA abnormalities which differ from sunlight-induced skin cancers.

Brain Tumours

The age-adjusted incidence of primary tumours of the central nervous system (CNS) (particularly astrocytomas, including the rapidly progressive glioblastomamulti form as well as the benign meningiomas) appears to have increased by 50–100 per cent over the past several decades, with greatest increase among the elderly (Davis et al., 1996). Studies also show increased occurrence of high-grade neuroepithelial tumours, lymphoma, and other primary CNS tumours of 5–13 per cent (Fleury et al., 1997) per annum in the elderly. Some observers attribute the apparent increase to the availability of computerized tomography (Christense et al., 1995), but disproportionate increase in certain histologic types (Werner et al., 1995), parallel increases in mortality (Shugg et al., 1994) and studies that show diagnostic imaging only contributes about 20 per cent to case ascertainment, all suggest the rise is probably real (Desmeules et al., 1992). Several studies of workers in farming (Khuder et al., 1998), gardening and orchard work (Littorin et al., 1993), pesticide application, (Figa-Talamanca et al., 1993) and golf-course superintendence (Kross et al., 1996) show increased risk for primary tumours of the brain. Research analysing risk of brain cancer among many occupational groups indicates that workers in occupations likely to involve pesticide exposure, heighten their liability to brain tumours (Rodval et al., 1996).

Respiratory Tract Cancer

Modest increase in cancers of the nose and nasal cavity is reported among workers exposed to phenoxy herbicides and chlorophenols. A greater than two-fold increase in lung cancer (adjusted for smoking) has been observed among structural pest-control workers (Pestaori et al., 1994). Excess cancer of the sinonasal cavities and lungs has been found among women working in agricultural settings.

Gastrointestinal Cancers

Gastric cancer has been associated with work as a farmer (Forastriere, 1993) as has colorectal cancer. In one retrospective cohort study, colorectal cancer specifically correlated with working in a plant that manufactured the herbicide alachlor. For all exposed workers, risk for developing leukemia or colorectal cancer was 50 per cent higher than for a comparable non-exposed population, while incidence of colorectal cancer among workers with five or more years of the highest alachlor exposure was more than five times greater (Leet et al., 1996). Research on workers in plants that manufacture organochlorines shows a nearly four-fold increased risk from exposure to chlordane, heptachlor, endrin, aldrin, and dieldrin (Brown, 1992). These pesticides are no longer used in the US, but persist in the environment-including termite-protected homes, so exposure may still occur. A number of studies implicate pesticides in pancreatic cancer. They show that occupational pesticide-exposure increases the risk of pancreatic cancer (Kauppinen et al., 1995). Workers exposed to DDT and related compounds suffer more than a seven-fold increased incidence of pancreatic cancer compared with non-exposed workers (Garabrant et al., 1992). In short, organochlorine exposure appears to be consistently linked with a variety of gastrointestinal malignancies.

Urinary Tract Cancer

The US has recently experienced increased incidence of and mortality from renal cancers. According to the Surveillance, Epidemiology and End Results (SEER) national cancer monitoring programme, the last 25 years have witnessed dramatic increases in disease and death from kidney cancer among black and white Americans of both sexes. During the last 20 years, white men saw increased incidence at 3.1 per cent per year; white women at 3.9 per cent; and African-American men and women, the steepest at 3.9 per cent and 4.3 per cent (Chow et al., 1999). Occupational exposure to pesticides (work in agriculture) has been correlated with increased risk for kidney cancer (or hypernephroma) (Mellemgaard et al., 1994). One study shows specific risk associated with pentachlorophenol (Ramlow et al., 1996). Among women occupationally exposed to pesticides, one study observed increased incidence of bladder cancer.

Testicular Cancer

Testicular cancer is another malignancy rising in occurrence for the last several decades in virtually all developed nations. Annual incidence increases range from 2.3 per cent to 5.2 per cent in Europe since the 1940s (Bergstrom et al., 1996). In Miyagi, Japan, growth is among the highest, with 6.6 per cent per annum (Forman and Moller, 1994). US data suggest similar trends: The nation's oldest on-going statewide tumor registry finds a mean annual increase in testicular cancer incidence of more than 5.5 per cent over the last 60 years (Zheng et al., 1996). Studies of offspring of parents who work in agricultural activities reveal higher rates of testicular cancer, with occurrence manifesting in childhood as well as young adulthood (Kristensen et al., 1996). Another study shows excess risk of testicular cancer among workers exposed to phenoxyherbicides and chlorophenols.

Breast Cancer

Age-adjusted incidence of breast cancer in industrialized countries has increased 1–2 per cent per year for several decades, both before and after introduction of mammography (Harmer et al., 1999; Tuliniu et al., 1999). This observation suggests environmental factors may play a role in this common disease. Recent years have witnessed great controversy over the possibility of attributing increased breast cancer incidence to hormonally active environmental contaminants, including some pesticides. The organochlorines have received special attention due to their estrogenic effects in vitro, lab animals, and wildlife. A case-control study of postmenopausal breast cancer measured serum levels of certain organochlorine compounds (DDE, hexachloro benzene, mirex, and several polychlorinated biphenyls or PCBs). Some increased risk appeared for women with certain types of PCBs and mirex detectable in their serum, but this effect was predominantly restricted top ostmenopausal women who had never breast-fed (Moysich et al., 1998). Another case-control study analysed breast tissue from patients with invasive cancer for the presence of organochlorines and compared it with control measurements from women with benign breast biopsies. Some, but not all, classes of PCBs were associated with breast cancer, especially among postmenopausal women with estrogen-receptor positive tumours. Hexachlorobenzene levels were also associated with increased risk of malignancy (Liljegren et al., 1998). In an ecological study of breast cancer incidence in an agricultural district heavily contaminated with organochlorine and triazine herbicides, a very modest but statistically significant increased risk of breast cancer is evident (Kettles et al., 1997).

List of pesticides which are highly carcinogenic are discussed in Table 3.1.

Table 3.1

Carcinogenic Pesticides

Chemical Name	*Chemical Use*
Arsenic acid	Herbicide
Arsenic pentoxide	Insecticide, wood preservative
Arsenic trioxide	Rodenticide
Carcodylic acid	Herbicide, defoliant
Captan	Fungicide
Chlorothalonil	Fungicide
Chromic acid	Wood preservative
Creosote	Wood preservative
Daminozide	Plant growth regulator
Ddvp	Insecticide
Dipropyl isocinchomeronate	Insecticide

Cont'd...

...Cont'd

Diuron	Herbicide
Ethoprop	Insecticide
Ethylene sodium	Fumigant
Fenoxycarb	Insecticide
Folpet	Fungicide
Formaldehyde	Microbiocide
Iprodione	Fungicide
Lindane	Insecticide
Mancozeb	Fungicide
Maneb	Fungicide
Metam-sodium	Fumigant
Metiram	Fungicide
Ortho-phenylphenol	Microbiocide
Ortho-phenylphenol, Sodium salt	Microbiocide
Oxadiazon	Herbicide
Oxythioquinon	Insecticide, Fungicide, Fumigant
Para-dichlorobenzene	Insecticide
Pentachlorophenol	Wood preservative
Potassium dichromate	Wood preservative
Propargite	Insecticide
Propoxur	Insecticide
Propylene oxide	Fumigant
Propyzamide	Herbicide
Pyrethrins	Insecticide
S.S.S.-tributyl phosphorotrithioate	Defoliant
Silica aeroge	Insecticide
Sodium dichromate	Wood preservative
Thiodicarb	Insecticide
Thiophanate-methyl	Fungicide
Trichlorfon	Insecticide
Vinclozolin	Fungicide

Source: Pesticides listed as known, likely, or probable carcinogens by US EPA Office of Pesticides Programs.

Pesticides and Respiratory Diseases

Acute organophosphate or N-methyl carbamate overexposure is well known to cause cholinesterase inhibition, resulting in broncho constriction, increased airway secretions, and respiratory distress. A few pesticides are known sensitizers and can result in allergic reactions including asthma (Honda et al., 1992). An association between low-level pesticide exposure and asthma is controversial and confounded by the fact that animal, plant, and other antigens cannot be completely ruled out. A few studies report other respiratory effects from pesticides, including pulmonary hemosiderosis, pneumonia-like infiltrates, chronic bronchitis, pulmonary fibrosis, Wegener's granulomatosis, and respiratory muscle impairment (Duna et al., 1998). The main target organ for the herbicide paraquat is the lung. This pesticide is selectively taken up by the lung from peripheral blood, and causes oxidative damage presenting as acute pulmonary edema and haemorrhage or as delayed pulmonary fibrosis. Respiratory failure has occurred following exclusively dermal exposure to this chemical.

Case reports and specific bronchial-challenge testing link several pesticides with occupational asthma. These pesticides include captafol, sulphur (Freedman 1980), pyrethrins and pyrethroids (Box and Lee, 1996), tetrachloroisophthalonitrile and several organophosphate and N-methyl carbamate insecticides that appear to have a methacholine-like effect on the lung (Weiner, 1961). Plantation workers in India showed a potential association between pesticide exposure and respiratory impairment. Although overall prevalence of asthma was lower among workers than among controls (perhaps due to the well-known 'healthy worker effect,' in which the working population, on average, enjoys a better health status than the overall population) (Li and Sung, 1999), the pesticide-exposed workers revealed an exposure-related increase in both obstructive and restrictive deficits on pulmonary function testing (Rastogi et al., 1989).

A cross-sectional study of nearly two thousand farmers in Saskatchewan revealed a significant association between physicians diagnosed asthma and reported use of cholinesterase inhibiting pesticides. Vineyard and orchard workers in Eastern Europe had significantly higher overall prevalence of dyspnea, chest tightness, chronic cough, and chronic phlegm compared with non-pesticide-exposed controls. Among both smoking and non-smoking workers employed for greater than ten years, FEV1, FEF25, and FEF50 were significantly reduced. Exposed workers also had significantly reduced FVC compared to controls. In a survey of about 200 Danish fruit-growers, individuals reported using an average of 13 different pesticides. The most commonly used pesticides comprised captan, paraquat, parathion, azinphos-methyl, diquat, amitrol, benomyl, and simazine. Approximately 40 per cent of the growers reported at least one significant respiratory symptom in connection with pesticide spraying, and nearly 20 per cent had diminished peak flow. These findings were more common among workers who did not wear respiratory protection when applying pesticides. X-ray revealed pulmonary infiltrates or fibrotic changes in nearly one quarter of the subjects.

Pesticides and Neurological Defects

Pesticides have been shown to affect both the central nervous system (CNS), and the peripheral nervous system (PNS) in animals and humans via a variety of mechanisms. The effects of neurotoxic pesticides can be assessed by measuring changes in neurochemistry, neuropathology and behaviour, including subtle effects on visuospatial function, concentration, reaction-time, learning, and short-term memory (Engel et al., 1998). Certain pesticides, for example, the organophosphates and N-methyl carbamates, are designed specifically to damage neurological function in insects and are neurotoxic in humans because of similarities in nervous system function between insects and humans. Human neurotoxic effects may be acute, may represent the chronic sequel of an acute poisoning, or may result from chronic exposures in the absence of an acute episode of poisoning (Table 3.2) (Evangelista et al., 1996).

Neurotoxicity

Neurotoxicity can result from high-level exposure to most types of pesticides, including organophosphates (OPs), carbamates, organochlorines, fungicides, and fumigants (Keifer and Mahurin, 1997), but only OPs have been studied in detail (He, 2000). The immediate response to OPs can occur within minutes. Mild cases display symptoms including headache, dizziness, nausea, vomiting, pupillary constriction, and excessive sweating, tearing, and salivation. More severe cases develop muscle weakness and muscle twitches, changes in heart rate, and bronchospasm and can progress to convulsions and coma. These symptoms are a consequence of overstimulation of postsynaptic cholinergic receptors following inhibition of acetyl cholinesterase by OPs. An intermediate syndrome, occurring one to four days after exposure, is characterized by muscle weakness and can be fatal if respiratory muscles are affected. Two to five weeks after exposure, some patients develop organophosphate-induced delayed polyneuropathy (OPIDP), a well-characterized syndrome involving sensory abnormalities, muscle cramps, weakness, and even paralysis, primarily in the legs. These symptoms are a consequence of axonal death following OP inhibition of a neural enzyme called neuropathy target esterase and may be irreversible. Several studies have shown that OP poisoning has long-term sequelae in addition to OPIDP. Studies of individuals with a history of pesticide poisoning, either farm workers (Ohayo et al., 2000) or from the general population (Steenland et al., 1994), have found that increased symptom prevalence, deficits in cognitive and psychomotor function, decreased vibration sensitivity, and impaired nerve conduction can occur long after the immediate episode is resolved. In some cases effects were observed 10 or more years after poisoning, which suggests that the residual damage is permanent.

Mood and Affect

Pesticide exposure is also associated with changes in mood and affect. Workers exposed to OPs (Steenland et al., 2000) or DDT (Van et al., 2001) reported higher levels of tension, anger, and depression. Farmers who applied pesticides reported more tension during the application season than off-season (Stokes et al., 1995), but this is not necessarily related to pesticide use. Farmers who reported a history of pesticide-related illness were more likely to score high on a depression scale (CES-D) than those who did not (Stallones and Beseler, 2002).

Parkinson's Disease

There is increasing evidence that a high proportion of Parkinson's Disease (PD) may be associated with environmental factors (Langston, 1998). Specific pesticides and pesticide classes implicated in PD include paraquat, the organophosphates, dieldrin, and the manganese-based fungicides maneb and mancozeb (Fleming et al., 1994). The designer heroin-like drug MPTP, known to cause a Parkinsonian syndrome in addicts via the neurotoxic effect of its major metabolite, is chemically related to the herbicide paraquat (Lewin, 1985). Several population-based case control studies identify a four-fold increased likelihood of past herbicide exposure among patients with PD, and a 3–4-fold increased likelihood of prior exposure to insecticides (Veldman et al., 1998). In particular, higher than expected rates of certain glutathione transferase polymorphisms, the slow acetylator genotype of N-acetyltransferase–2, and the slow 4-hydroxylation of debrisoquine (the CYP 2D6 29B+ allele) have all been reported in patients with Parkinson's disease (Hubble et al., 1998). These genetic variants may increase risk from environmental exposure by slowing detoxification of exogenous compounds.

Seizures

Many pesticides are known to increase CNS excitability and to produce seizures with acute high-dose exposure (O'Malley, 1997). Low doses repeated three times a week for ten weeks of the pesticide lindane (used to treat head lice) resulted in enhanced myoclonic jerks and seizures at normally subconvulsant doses. Other organochlorine pesticides, such as endosulfan and dieldrin, are reported to have similar effects (Gilbert, 1995).

Table 3.2

Chronic or Delayed Neurotoxic Effects of Pesticides

Pesticide Category	*Effects on Central Nervous System*	*Effect on Peripheral Nervous System*
Organophosphates e.g. malathion, chlorpyrifos	Cognitive, affective and perspective effects	Sensorimotor neuropathy, intermediate syndrome
Carbamates e.g. carbaryl	Memory deficits, visual impairment	Sensorimotor neuropathy
Organochlorines e.g. kepone	Impairment of cognitive function and personality, seizure kindling	Tremor (kepone shakes)
Metals e.g. zinc phosphide, lead arsenate	Impaired visuopatial abilities, deficits in short-term verbal memory	Painful, burning dysesthesias
Fumigants e.g. methyl bromide, dichloropropene	Cognitive impairment, mood changes, difficulty in concentrating	Loss of reflexes and distal motor strength
Fungicides e.g. dithiocarbamates	Reduction of physiologic respiratory arrhythmia, possibly parkinson's	Reduced nerve conduction
Pyrethroids e.g. cypermethrin, fenvalerate	Reduction of spontaneous motor activity	Cutaneous paresthesia, numbness
Rodenticides e.g. vacor	Minimal data on cognitive impairment	Autonomic incompetence

Reproductive and Developmental Effects of Pesticides

Pesticides may affect human reproduction by direct toxicity to the reproductive organs or by interference with hormonal function (Garcia, 1998). Effects of pesticides on reproduction may include menstrual abnormalities, male or female infertility, or hormonal disturbances. The developing fetus and infant are disproportionately susceptible to the health effects of pesticides. Developmental toxicity of pesticides may result in spontaneous abortion, growth retardation, structural birth defects, or functional deficits.

Growth Retardation and Spontaneous Abortion

Numerous studies report an increased rate of spontaneous abortions and stillbirths among female agricultural workers. These studies are limited by potential recall bias and by difficulties in exposure assessment, since workers are exposed to a complex mixture of chemicals and doses are unknown. Some studies of wives of agricultural workers also show an increased risk of spontaneous abortion and stillbirth (Rupa et al., 1991). A California study demonstrated an association between pesticide exposure at work or in the home and stillbirths, particularly those with congenital anomalies. Elevated risks ranged from a 70 per cent increased risk of stillbirth for home exposure to pesticides, to a 240 per cent increased risk for occupational exposure. Higher levels of organochlorine pesticides have been found in abortuses and preterm infants than in full-term babies (Saxena et al., 1981). Women living in communities supplied with drinking water contaminated by a variety of herbicides, including atrazine, cyanazine, and metolachlor, had an 80 per cent increased risk of intrauterine growth retardation compared with similar communities with uncontaminated water (Munger et al., 1997).

Birth Defects

Numerous epidemiological studies and case reports associate pesticide exposure at work or home with increased risk of various types of congenital malformations (McDonald et al., 1988). Particular birth defects associated with pesticides include:

- *Cleft lip and palate:* a doubling of risk with exposure during the first trimester (Nurminen et al., 1995).
- *Limb defects:* a three-four-fold increased risk for garden or workplace exposure, and greater than doubling of risk with household exposure, particularly if pesticides were applied by a professional pest eradication service (Kristensen et al., 1997).
- *Cardiovascular malformations, particularly Total Anomalous Pulmonary Venous Return:* a two to three times greater risk found in the Baltimore-Washington Infant Study.
- *Spina bifida and hydrocephaly:* a 2.7- and 3.5 times increased risk respectively, in one study, and a 50 per cent increased risk with residence within a quarter-mile of an agricultural field in another.

- *Cryptorchidism and hypospadias:* a tow to three-fold greater rate of orchidopexy in highly agricultural areas; a 50 per cent increase in hypospadias also reported (Garcia et al., 1996).

Disruption of Hormone Function

Various pesticides mimic estrogen, while others block androgens or thyroid hormone. Around 48 Estrogenic pesticides that have been studied in some detail include numerous banned and still used organo-chlorine pesticides, such as DDT, chlordecone, dicofol, methoxychlor, endosulfan, and lindane (Bason and Colborn, 1998). Fungicides such as vinclozolin and iprodione are antiandrogens (Gray et al., 1999). In addition, some triazine herbicides such as atrazine interfere with estrogen via indirect pathways. Male pesticide factory workers in China exposed to the organophosphate pesticides ethyl parathion and methamidophos had significant abnormalities in their reproductive hormone profiles. Increased pesticide exposure correlated positively with serum LH and FSH levels, and negatively with serum testosterone. In addition, workers with higher exposure tended to show greater risk of abnormal semen parameters (Padungtod et al., 1998). Workers applying ethylene bis dithiocarbamate fungicides (such as Maneb or Zineb) in Mexico developed elevated levels of TSH without changes in thyroid hormone levels. Although findings were subclinical in these healthy adult males, they could be relevant to a developing fetus where a pregnant woman is exposed. In the foetus or neonate, disruption of endocrine homeostasis can result in permanent alterations in sexual development, whereas disturbance in adulthood is less likely to create lasting health effects.

Methods to Detect Pesticides

A wide range of methodologies exist to identify possible exposure to pesticides. When identification is necessary due to poisoning of a patient attended to in the clinic, the general procedures include anamnesis, physical examination, evaluation of clinical signs, and diagnostic and toxicological analysis. If the investigation aims to qualify and/or quantify a possible pesticide, it is generally necessary to collect a sample and analyse it for the presence of pesticides and/or metabolites in biological samples (blood, liver, stomach contents) and/or the environment (air, water, and ground). Selection of the test will depend on the purpose of the analysis. It is also essential to consider the financial costs of a method. Simpler tests are still important – apart from been inexpensive, many offer high sensitivity, specificity, precision, and accuracy, all of which are factors that are crucial for reliable analysis (Niessen, 2012). Prior to analysing pesticides samples, analysts have to go through similar steps: definitions of the analytical problem (target analyte and its properties), choice of detection methods (immunoassays spectrometry), sampling (how to collect and store the sample), sample preparation (solubilization, extraction, concentration, and separation), calibration (qualification and/or quantification of the analyte), calculation and evaluation of the results, and actions to complete the analysis (Wen et al., 2014).

Sample storage for long periods should ensure that no sample degradation or external contamination occurs. Well-sealed containers stored under refrigeration and protected from light are mandatory. To avoid any type of external interference during analysis, none of the employed materials should modify or degrade the pesticide in the sample. The analysis of pesticides, mainly in water, ambient air and soil sediments, often requires a purification step to clean the sample and pre-concentrate the analytes, to improve the quality of the analytical results. The extraction process is a key analytical step – it extracts the desirable compounds for further separation and characterization. Liquid-liquid extraction and pre-concentration procedures, such as solid-phase extraction and solid-phase micro extraction, are the most commonly used methods, but other extraction methods are also applicable depending on the objective (Martins et al., 2013). Extraction of residues from the sample matrix demands appropriate solvents for maximum extraction efficiency and minimal co-extraction of interfering substances. The extraction solvents must be highly pure. Blank tests help to prove that the matrix does not interfere in the analyses. After extraction, a purification step removes the interfering substance with minimal loss of the analyte. The final solution should include an appropriate solvent for analyte determination by the selected method (Santaladchaiyakit et al., 2012).

Below is a didactic description of the main separation and detection methods:

Physicochemical Methods

Gas Chromatography (GC), Liquid Chromatography (LC), and Capillary Electrophoresis (CE) constitute physicochemical separation methods.

When the analysed pesticide is volatile or semi-volatile, GC still is the method of choice: it offers higher resolution and lower detection limits. GC is usually associated with multiple detectors whose choice will depend on the characteristics of the target analytes. GC is based on sample volatilization and introduction into a chromatographic column coated or packed with a solid or liquid stationary phase. A gaseous mobile phase elutes the analyte; this phase is inert, and does not interact with the analyte. The carrier gases should be pure and chemically inert, too, and the choice will depend on the detector. The commonest carrier gases are helium, argon, nitrogen, carbon dioxide, and hydrogen (Skoog and Leary, 1992).

LC has emerged as a great separation tool. It allows for effective separation of nonvolatile and thermally unstable pesticides that are incompatible with GC. During LC, extracts pass through multiple adsorbent columns that can discriminate between the components of the matrix and target analyte. The degree of selectivity will vary according to the adsorbent present in the column (alumina, silica gel, or florisil), mesh size, and activity levels. Columns can be used separately or in combination (Andreu and Pico, 2012).

CE is a powerful tool to separate and identify a wide range of molecules. EC provides high resolution, and large separation efficiency. It requires small sample size and low solvent consumption analyses is faster and operational costs are low (Assuncao et al., 2008).

An ideal detector should ensure adequate sensitivity, good stability and reproducibility and linear response to various concentrations of the analytes. It should also operate in a wide range of temperature, have reduced response time (independent of the flow), and be easy to handle. The detector response should be equivalent for all the analytes or selective to certain classes of compounds. Ultimately, the detector should not destroy the sample. Unfortunately, a switch that exhibits all these characteristics does not exist, so it is necessary to select the detector according to the desired goal. Several types of detectors are commercially available. They can come coupled to the separation device. These detectors use photometric or fluorimetric methods, thermal conductivity, diode array detection, electrons capture, atomic absorption, or pesticides mass/charge evaluation. The latter method is currently in evidence because it is highly sensitive, offers autonomy and performs a variety of functions. Electron capture and mass spectrometry are most often used to detect pesticides.

The electron capture detector (ECD) is usually employed to search for organic pesticides, because it is highly sensitive and selective toward molecules containing electronegative functional groups. It also detects masses in the order of pictograms and can analyse traces of pesticides. However, ED cannot detect compounds with low electron affinity. Its excellent properties are useful for analysis of pesticides in both the environmental area and hospitals. A detector called μECD is also available in the market. It is advantageous over ECD in term of sensitivity, stability, and robustness (Poole, 2013).

Mass spectrometry (MS) is based on the ionization, acceleration, and separation of the generated molecules and ions according their mass/charge (m/z) ratio. This provides a typical spectrum that gives the relative mass abundance of the different ionic species as a function of m/z, which permits unambiguous identification of molecules. Mass spectrometry is a confirmation technique that is less subject to misunderstanding. Nevertheless, it has a drawback – it destroys the analyte (Di Stefano et al., 2012). As mentioned previously, the choice of method will depend on the case. LC-MS and GC-MS are the methods that generally separate and detect pesticides most suitable. These methods play a very important role in the analysis of pesticides and related compounds and are applicable in several areas like environmental analysis, food safety, and occupational toxicology, among others. Because they can serve various purposes, these methods also help to detect compounds in different samples, such as water, soil, sediment, sludge, vegetables and fruits, and animals and humans tissues and fluids. Obviously, method will be based on the needs and characteristics of the target pesticide, and each sample will have their own features, which will depend on their physicochemical properties.

Biological

Chemical analysis of isolated compounds is commonly used to monitor environmental pollution, but such analyses can be limited and expensive and cannot indicate the biological effects. In contrast, biological tests indicate the toxicity of a wide range of compounds or environmental samples, and are therefore essential to determine the environmental impact of the presence of these chemicals (Oliveira et al., 2012). Immunoassays and biosensors are methods related to the biological factor. Immunoassays are a powerful tool in clinical laboratories and one of the most widely applied analytical techniques. The reagents kits and the equipment necessary to perform immunoassays are commercially available and rely on fluorescent, chemiluminescent or other detection methods. Immunoassays can detect a wide range of compounds including drugs, proteins, and hormones; they can also identify and quantify the presence of pesticides residues in various samples such as natural water, food and blood, among others.

Regarding biosensors, organisms such as *Drosophila melanogaster* fly species may aid the detection of pesticides in food samples and other matrixes such as water, soil, plants and animal tissue. This test model is advantageous, because these insects have low tolerance to toxic substances with insecticidal character, besides being experimental models of easy creation, manipulation and maintenance. In addition, they require few financial resources and can remain under laboratory conditions. However, this method only serves to detect the presence of pesticides, but it cannot identify the detected compound. Therefore, after using this probe, the analyst has to employ a chromatographic, for example, to identify the group of pesticides in that sample.

Alternative to Harmful Chemical Pesticides

For the average gardener, the use of organic pesticides can keep a healthy balance in the soil. Many organic pesticides are made of minerals or other plant materials that will keep pests at bay and break down quickly in the soil (Chaturvedi et al., 2013).

Examples of some common organic pesticides include the following:

- *Cayenne pepper spray:* Can be sprayed on the leaves of plants to deter harmful insect.
- *Soap spray:* Also sprayed on plants to get rid of aphids.
- *Tobacco powder:* A spray can be made from the finely ground tobacco leaves and water. It is used to kill sucking insects on plants such as aphids, thrips and spider mites.
- *Neem:* Derived from the neem tree. Used to control Gypsy moths, leaf miners, mealy bugs, whiteflies and caterpillars.
- *Sabadilla:* Derived from the sabadilla lily. Used to control caterpillars, leaf hoppers, stink bugs and squash bugs.

Conclusion

For decades people have believed that harmful chemical pesticides were the only true way to rid gardens and crop fields from pests. Pesticides have contaminated almost every part of our environment, as pesticide residue are found in soil, air and in surface and groundwater across the nation and urban pesticide use contributes to the problem. Pesticide contamination poses significant risks to the environment and non-target organisms ranging from beneficial soil micro-organisms to insects, plants, fish and birds. Contrary to common misconceptions, even herbicides can cause harm to the environment. In fact, weed killers can be especially problematic because they are used in relatively large volumes. Soil pollution, air pollution has occurred from the use of pesticides and it takes years and sometimes decades for some of these chemicals to break down. Luckily there are many organic chemicals that are just as effective. The effects of pesticides on soil micro-organisms are less invasive when organic pesticides are used.

People need to break the habit of using harmful pesticides and switch to rising organic ones that break down quickly in the sunlight and in the soil. The faster a chemical breaks down, the sooner the soil can return to a healthy state. Most organic pesticides are also safe to use around people and pets. They can easily be washed from fruits and vegetables making them healthier for us and our family to eat.

References

Aktar, M.W. (2008). Impact of Pesticides Use in Agriculture: Their Benefits and Hazards. *Interdisciplinary Toxicology*. 2(1): 1–12.

Andreu, V. and Pico, Y. (2012). Determination of Currently used Pesticides in Biota. *Analytical and Bio Analytical Chemistry*. 404(9): 2659–81.

Assunçao, N.A., Bechara, E.J.H., Simionato, A.V.C., Tavares, M.F.G. and Carrilho, E. (2008). Capillary Electrophoresis Coupled to Mass Spectrometry (CE-MS): Twenty Years of Development. *Química Nova*. 31: 2124–2133.

Bason, C.W. and Colborn, T. (1998). US Application and Distribution of Pesticides and Industrial Chemicals Capable of Disrupting Endocrine and Immune Systems. *Environmental Toxicology and Occupational Medicine*. 7: 147–56.

Bergstrom, R., Adami, H.O. and Mohne, M. (1996). Increase in Testicular Cancer Incidence in Six European Countries: A Birth Cohort Phenomenon. *Journal of the National Cancer Institute*. 88: 727–33.

Bhardwaj, T. and Sharma, J.P. (2013). Impact of Pesticides Application in Agriculture Industry: An Indian Scenario. *International Journal of Agriculture and Food Science Technology*. 4(8): 817–822.

Blair, A., Cantor, K.P. and Zahm, S.H. (1998). Non-Hodgkin's Lymphoma and Agricultural Use of the Insecticide Lindane. *American Journal of Industrial Medicine*. 33: 82–87.

Box, S.A. and Lee, M.R. (1996). A Systemic Reaction Following Exposure to a Pyrethroid Insecticide. *Human and Experimental Toxicology*. 15: 389–90.

Brown, D.P. (1992). Mortality of Workers Employed at Organochlorine Pesticide Manufacturing Plants: An Update. *Scandinavian Journal of Work, Environmental and Health*. 18: 155–61.

Cartwright, R., Brincker, H. and Carli, P.M. (1999). The Rise in Incidence of Lymphomas in Europe 1985–1992. *European Journal of Cancer*. 35: 627–33.

Cantor, K.P., Blair, A. and Everett, G. (1992). Pesticides and other Agricultural Risk Factors for Non-Hodgkin's Lymphoma Among Men in Lowa and Minnesota. *Cancer Research*. 52: 2447–55.

Chaturvedi, M., Sharma, C. and Chaturvedi, M. (2013). Effects of Pesticides on Human Beings and Farm Animals: A Sase Study. *Research Journal of Chemical and Environmental Sciences*. 1(3): 14–19.

Chow, W.H., Devesa, S.S., Warren, J.L. and Fraumen, J.F. (1999). Rising Incidence of Renal Cell Cancer in the United States. *Journal of the American Medical Association*. 281: 1628–31.

Christensen, J., Klarskov, H. and Raffin, E. (1995). Primary Intracranial and Intraspinal Neoplasms in Denmark 1943–1987. *Ugeskr Laeger*. 157: 5716–20.

Ciccone, G., Mirabelli, D. and Levis, A. (1993). Myeloid Leukemias and Myelodysplastic Syndromes: Chemical Exposure, Histologic sub-type and Cytogenetics in a Case-control Study. *Cancer Genetics and Cytogenetics*. 68: 135–39.

Cuzick, J. (1994). Multiple Myeloma. *Cancer Survival*. 19–20: 455–74.

Davis, F.G., Malinski, N. and Haenszel, W. (1996). Primary Brain Tumour Incidence Rates in Four United States Regions, 1985–1989: A Pilot Study. *Neuroepidemiology*. 15: 103–12.

Demers, P.A., Vaughan, T.L. and Koepsell, T.D. (1993). A Case-Control Study of Multiple Myeloma and Occupation. *American Journal of Industrial Medicine*. 23: 629–39.

Desmeules, M., Mikkelsen, T. and Mao, Y. (1992). Increasing Incidence of Primary Malignant Brain Tumours: Influence of Diagnostic Methods. *Journal of the National Cancer Institute*. 84: 442–45.

Devesa, S.S., Silverman, D.T. and Young, J.L. (1987). Cancer Incidence and Mortality Trends among Whites in the United States, 1947–84. *Journal of the National Cancer Institute*. 79: 701–70.

Devesa, S.S. and Fears, T. (1992). Non-Hodgkin's Lymphoma Time Trends: United States and International Data. *Cancer Research*. 52: 5432–40.

Di Stefano, V., Avellone, G., Bongiorno, D., Cunsolo, V., Muccilli, V., Sforza, S., Dossena, A., Drahos, L. and Vekey, K. (2012). Applications of Liquid Chromatography-Mass Spectrometry for Food Analysis. *Journal of Chromatography A*. 1259: 74–85.

Duna, G.F., Cotch, M.F. and Galperin, C.N. (1998). Wegener's Granulomatosis: Role of Environmental Exposures. *Clinical and Experimental Rheumatology*. 16: 669–74.

Engel, L.S., Keifer, M.C. and Checkoway, H. (1998). Neurophysiological Function in Farm Workers Exposed to Organophosphate Pesticides. *Archives of Environmental Health*. 53: 7–14.

Environmental Protection Agency (EPA). (2004). Overview of the Ecological Risk Assessment Processing the Office of Pesticide Programs, US, Environmental Protection Agency.

Eriksson, M. and Karlsson, M. (1992). Occupational and Other Environmental Factors and Multiple Myeloma: A Population Based Case-Control Study. *British Journal of Industrial Medicine.* 49: 95–103.

Evangelista, A.M., Duffard, D. and Duffard, R. (1996). Behavioural Toxicology, Risk Assessment, and Chlorinated Hydrocarbons. *Environmental Health Perspectives.* 104: 353–60.

Fenik, J., Tankiewicz, M. and Biziuk, M. (2011). Properties and Determinations of Pesticides in Fruits and Vegetables. *Trends in Analytical Chemistry.* 30(6): 814–826.

Figa-Talamanca, I., Mearell, I. and Valente, P. (1993). Mortality in a Cohort of Pesticide Applicators in an Urban Setting. *International Journal of Epidemiology.* 22: 674–76.

Fleming, L., Mann, J.B. and Bean, J. (1994). Parkinson's Disease and Brain Levels of Organochlorine Pesticides. *Annals of Neurology.* 36: 100–3.

Fleury, A., Menengoz, F. and Grosclaude, P. (1997). Descriptive Epidemiology of Cerebral Gliomas in France. *Cancer.* 79: 1195–1202.

Forman, D. and Moller, H. (1994). Testicular Cancer. *Cancer Survival.* 19–20: 323–41.

Forastriere, F., Quercia, A. and Miceli, M. (1993). Cancer Among Farmers in Central Italy. *Scandinavian Journal of Work, Environmental and Health.* 19: 382–89.

Freedman, B.J. (1980). Sulphur Dioxide in Foods and Beverages: Its Use As a Preservative and its Effect on Asthma. *British Journal of Diseases of the Chest.* 74: 128–34.

Garcia, A.M. (1998). Occupational Exposure to Pesticides and Congenital Malformations: A Review of Mechanisms, Methods, Results. *American Journal of Industrial Medicine.* 33: 232–40.

Garcia-Rodriguez, J., Garcia-Martin, M. and Nogueras-Ocana, M. (1996). Exposure to Pesticides and Cryptorchidism: Geographical Evidence of a Possible Association. *Environmental Health Perspectives.* 104: 1090–95.

Garabrant, D.H., Held, J. and Langholz, B. (1992). DDT and Related Compounds and Risk of Pancreatic Cancer. *Journal of the National Cancer Institute.* 84: 764–71.

Gilbert, M.E. (1995). Repeated Exposure to Lindane Leads to Behavioural Sensitization and Facilitates Electrical Kindling. *Neurotoxicology Teratology.* 17: 131–41.

Gray, L.E., Ostby, J., Monosson, E. and Kelce, W.R. (1999). Environmental Anti-Androgens: Low Doses of the Fungicide Vinclozolin Alter Sexual Differentiation of the Male Rat. *Toxicology and Industrial Health.* 15: 48–64.

Hansen, E.S., Hasle, H., and Lande, F. (1992). A Cohort Study on Cancer Incidence Among Danish Gardeners. *American Journal of Industrial Medicine*. 21: 651–60.

Hardell, L. and Eriksson, M. (1999). A Case-control Study of Non-Hodgkin's Lymphoma and Exposure to Pesticides. *Cancer*. 85: 1353–60.

Harmer, C., Staples, M. and Kavanaugh, A.M. (1999). Evaluation of Breast Cancer Screening Incidence: Is the Increase Due Entirely to Mammographic Screening? *Cancer Causes and Control*. 10(5): 333–37.

Hartge, P. and Devesa, S.S. (1992). Quantification of the Impact of Known Risk Factors on Time Trends in Non-Hodgkin's Lymphoma Incidence. *Cancer Research*. 52: 5566.

Hartge, P., Devesa, S.S. and Fraumeni, J.F. (1994). Hodgkin's and Non-Hodgkin's Lymphomas. *Cancer Survival*. 19–20: 423–53.

He, F. (2000). Neurotoxic Effects of Insecticides–current and Future Research: A Review. *Neurotoxicology*. 21: 829–35.

Hernandez, A.F., Parron, T., Tsatsakis, A.M., Requena, M., Alarcon, R. and Lopez-Guarnido, O. (2013). Toxic Effects of Pesticide Mixtures at a Molecular Level: Their Relevance to Human Health. *Toxicology*. 307: 136–145.

Hjalgrim, H., Frisch, M., Begtrup, K. and Melbye, M. (1996). Recent Increase in the Incidence of Non-Hodgkin's Lymphoma Among Young Men and Women in Denmark. *British Journal of Cancer*. 73: 951–54.

Honda, I., Kohrogi, H. and Araki, S. (1992). Occupational Asthma Induced by the Fungicide Tetrachloroisophthalonitrile. *Thorax*. 47: 760–61.

Hoppin, J.A., Tolbert, P.E. and Herrick, R.F. (1998). Occupational Chlorophenol Exposure and Soft Tissue Sarcoma Risk among Men Aged 30–60 Years. *American Journal of Epidemiology*. 148: 693–703.

Hsieh, L.L., Chen, H.J. and Hsieh, J.T. (1994). Arsenic-related Bowen's Disease and Paraquat-related Skin Cancer Lesions Show no Detectable Ras and p53 gene Alterations. *Cancer Letters*. 86: 59–65.

Hubble, J.P., Kurth, J.H. and Glatt, S.L. (1998). Gene-toxin Interaction as a Putative Risk Factor for Parkinson's Disease with Dementia. *Neuroepidemiology*. 17: 96–104.

Kauppinen, T., Partanen, T., Degerth, R. and Ojajarvi, A. (1995). Pancreatic Cancer and Occupational Exposures. *Epidemiology*. 6 : 498–502.

Keifer, M, Mahurin, R. 1997. Chronic Neurologic Effects of Pesticide Overexposure. *Occupational Medicine*. 12: 291–304.

Kettles, M.K., Browning, S.R., Prince, T.S. and Horstman, S.W. (1997). Triazine Herbicide Exposure and Breast Cancer Incidence: An Ecologic Study of Kentucky Counties. *Environmental Health Perspectives*. 105: 1222–27.

Khuder, S.A. and Mutgi, A.B. (1997). Meta-analysis of Multiple Myeloma and Farming. *American Journal of Industrial Medicine*. 32: 510–16.

Khuder, S.A., Mutgi, A.B., and Schaub, E.A. (1998). Meta-analyses of Brain Cancer and Farming. *American Journal of Industrial Medicine*. 34: 252–60.

Kogevinas, M., Kauppinen, T. and Winkelmann, R. (1995). Soft Tissue Sarcoma and Non-Hodgkin's Lymphoma in Workers Exposed to Phenoxyherbicides, Chlorophenols, and Dioxins: Two Nested Case-control Studies. *Epidemiology*. 6: 396–402.

Kristensen, P., Andersen, A. and Irgens, L.M. (1996). Cancer in Offspring of Parents Engaged in Agricultural Activities in Norway: Incidence and Risk Factors in the Farm Environment. *International Journal of Cancer*. 65: 39–50.

Kristensen, P., Andersen, A. and Irgens, L.M. (1996). Incidence and Risk Factors of Cancer Among Men and Women in Norwegian Agriculture. *Scandinavian Journal of Work, Environmental and Health*. 22: 14–26.

Kristensen, P., Irgens, L.M. and Andersen, A. (1997). Birth Defects Among Offspring of Norwegian Farmers, 1967–1991. *Epidemiology*. 8: 537–44.

Kross, B.C., Burmeister, L.F. and Ogilvie, L.K. (1996). Proportionate Mortality Study of Golf Course Superintendents. *American Journal of Industrial Medicine*. 29: 501–6.

Langston, J.W. (1998). Epidemiology versus Genetics in Parkinson's Disease: Progress in Resolving an Age-old Debate. *Annals of Neurology*. 44: S45–52.

Leet, T., Acquavella, J. and Lynch, C. (1996). Cancer Incidence Among Alachlor Manufacturing Workers. *American Journal of Industrial Medicine*. 30: 300–6.

Lewin, R. (1985). Parkinson's Disease: An Environmental Cause? *Science*. 229: 257–58.

Li, C.Y. and Sung, F.C. (1999). A Review of the Healthy Worker Effect in Occupational Epidemiology. *Occupational Medicine*. 49: 225–29.

Liljegren, G., Hardel, L. and Lindstrom, G. (1998). Case-control Study on Breast Cancer Concentrations of Congener Specific Polychlorinated Biphenyls, DDE and Hexachlorobenzene. *European Journal of Cancer Prevention*. 7: 135–40.

Littorin, M., Attewell, R. and Skervfving, S. (1993). Mortality and Tumour Morbidity among Swedish Market Gardeners and Orchardists. *International Archives of Occupational and Environmental Health*. 65: 163–69.

Mathur, S.C. (1999). Future of Indian Pesticides Industry in Next Millennium. *Pesticide Information*. 24(4): 9–23.

Martins, J.G., Amaya Chavez, A., Waliszewski, S.M., Colin Cruz, A. and Garcia Fabila, M.M. (2013). Extraction and Clean-up Methods for Organochlorine Pesticides Determination in Milk. *Chemosphere*. 92(3): 233–46.

Matthews, G., Bateman, R. and Miller, P. (2008). Pesticide Application Methods. John Wiley & Sons.

McDonald, A.D., McDonald, J.C. and Armstrong, B. (1988). Congenital Defects and Work in Pregnancy. *British Journal of Industrial Medicine.* 45: 581–88.

Mellemgaard, A., Engholm, G., McLaughlin, J.K. and Olsen, J.H. (1994). Occupational Risk Factors for Renal-cell Carcinoma in Denmark. *Scandinavian Journal of Work, Environmental and Health.* 20: 160–65.

Mostafalou, S. and Abdollahi, M. (2013). Pesticides and Human Chronic Diseases: Evidences, Mechanisms and Perspectives. *Toxicology and Applied Pharmacology.* 268: 157–177.

Moysich, K.B., Ambrosone, C.B. and Vena, J.E. (1998). Environmental Organochlorine Exposure and Postmenopausal Breast Cancer Risk. *Cancer Epidemiology Biomarkers Prevention.* 7: 181–88.

Munger, R.G., Isaacson, P. and Hu, S. (1997). Intrauterine Growth Retardation in Iowa Communities with Herbicide Contaminated Drinking Water Supplies. *Environmental Health Perspectives.* 105: 308–314.

Nanni, O., Falcini, F. and Buiatti, E. (1998). Multiple Myeloma and Work in Agriculture: Results of a Case-control Study in Forli, Italy. *Cancer Causes and Control.* 9: 277–83.

Niessen, W.M. (2012). Fragmentation of Toxicologically Relevant Drugs in Negative-ion Liquid Chromatography-tandem Mass Spectrometry. *Mass Spectrometry Reviews.* 31(6): 626–65.

Nurminen, T., Rantala, K., Kurppa, K. and Holmberg, P.C. (1995). Agricultural Work During Pregnancy and Selected Structural Malformations in Finland. *Epidemiology.* 6: 23–30.

O'Malley, M.A. (1997). Clinical Evaluation of Pesticide Exposure and Poisonings. *Lancet.* 349: 1161–66.

Ohayo-Mitoko, G.J., Kromhout, H, Simwa, J.M., Boleij, J.S. and Heederik, D. (2000). Self-reported Symptoms and Inhibition of Acetyl Cholinesterase Activity among Kenyan Agricultural Workers. *Occupational and Environmental Medicine.* 57: 195–200.

Oliveira, G.A.R., Lapuente, J., Leme, D.M., Ferraz, E.R.A., Meireles, G. and Oliveira, D.P. (2012). New Paradigms for the Environmental Assessment: An Eco-toxicological and Genetic Approach. In: *Advances in Environmental Research.* Nova Science Publishers.

Padungtod, C., Lasley, B.L. and Christiani, D.C. (1998). Reproductive Hormone Profile among Pesticide Factory Workers. *Journal of Occupational and Environmental Medicine.* 40: 1038–47.

Pasqualetti, P., Casale, R., Colantonio, D. and Collacciani, A. (1991). Occupational Risk for Hematological Malignancies. *American Journal of Hematology*. 38: 147–49.

Pesatori, A.C., Sontag, J.M. and Lubin, J.H. (1994). Cohort Mortality and Nested Case-control Study of Lung Cancer Among Structural Pest Control Workers in Florida. Cancer *Causes and Control*. 5: 310–18.

Pernille, A., Grandjean, P., Jorgensen, T., Brock, J.W. and Hartvig, H.B. (1998). Organochlorine Exposure and Risk of Breast Cancer. *Lancet*. 352: 1816–1820.

Poletta, G.L., Larriera, A., Kleinsorge, E. and Mudry, M.D. (2009). Genotoxicity of the Herbicide Formulation Roundup (glyphosate) in Broad-snouted Caiman (Caiman latirostris) Evidenced by the Comet Assay and the Micronucleus Test. *Mutation Research*. 672: 95–102.

Pollan, M., Lopez-Abente, G. and Pla-Mestre, R. (1993). Time Trends for Multiple Myeloma in Spain, 1957–1986. *International Journal of Epidemiology*. 22: 45–50.

Pollan, M., Lopez-Abente, G. and Moreno, C. (1998). Rising Incidence of Non-Hodgkin's Lymphoma in Spain: Analysis of Period and Cohort Effects. *Cancer Epidemiology Biomarkers Prevention*. 7: 621–25.

Poole, C.F. (2013). Derivatization Reactions for Use with the Electron-capture Detector. *Journal of Chromatography*. 1296: 15–24.

Pottern, L.M., Heineman, E.F. and Olsen, J.H. (1992). Multiple Myeloma Among Danish Women: Employment History and Workplace Exposures. *Cancer Causes and Control*. 3: 427–32.

Ramlow, J.M., Spadacene, N.W. and Hoag, S.R. (1996). Mortality in a Cohort of Pentachlorophenol Manufacturing Workers, 1940–1989. *American Journal of Industrial Medicine*. 30: 180–94.

Rastogi, S.K., Gupta, B.N. and Husain, T. (1989). Study of Respiratory Impairment among Pesticide Sprayers in Mango Plantations. *American Journal of Industrial Medicine*. 16: 529- 38.

Richardson, S., Zittoun, R. and Bastuji-Garin, S. (1992). Occupational Risk Factors for Acute Leukemia: A Case-control Study. *International Journal of Epidemiology*. 21: 1063–73.

Rodval, Y., Ahlbom, A., Spannare, B. and G. Nise, G. (1996). Glioma and Occupational Exposure in Sweden: A Case-control Study. *Journal of Occupational and Environmental Medicine*. 53: 526–37.

Rupa, D.S., Reddy, P.P. and Reddi, O.S. (1991). Reproductive Performance in a Population Exposed to Pesticides in Cotton Fields in India. *Environmental Research*. 55: 123–28.

Santaladchaiyakit, Y., Srijaranai, S. and Burakham, R. (2012). Methodological aspects of Sample Preparation for the Determination of Carbamate Residues: A Review. *Journal of Separation Science.* 35 (18): 2373–2389.

Saxena, M.C., Siddiqui, M.K. and Seth, T.D. (1981). Organochlorine Pesticides in Specimens from Women Undergoing Spontaneous Abortion, Premature or full-term Delivery. *Journal of Analytical Toxicology.* 5: 6–9.

Schreinemachers, D.M., J.P. Creason, J.P. and Garry V.F. (1999). Cancer Mortality in Agricultural Regions of Minnesota. *Environmental Health Perspectives.* 107: 205–11.

Shugg, D., Allen, B.J. and Blizzard, L. (1994). Brain Cancer Incidence and Mortality and Case Survival: Observations from Two Australian Cancer Registries. *International Journal of Cancer.* 59: 765–70.

Skoog, D. and Leary, J. (1992). *Principles of Instrumental Analysis*, (4th Ed.), Saunders College Publishing.

Steenland, K., Jenkins, B., Ames, R.G., O'Malley, M., Chrislip, D. and Russo, J. (1994). Chronic Neurological Sequelae to Organophosphate Pesticide Poisoning. *American Journal of Public Health.* 84: 731–36.

Steenland, K., Dick, R.B., Howell, R.J., Chrislip, D.W., Hines, C.J. and Reid, T.M. (2000). Neurologic Function among Termiticide Applicators Exposed to Chlorpyrifos. *Environmental Health Perspectives.* 108: 293–300.

Stokes, L., Stark, A., Marshall, E. and Narang, A. (1995). Neurotoxicity among Pesticide Applicators Exposed to Organophosphates. *Occupational and Environmental Medicine.* 52: 648–53.

Stallones, L. and Beseler, C. (2002). Pesticide Poisoning and Depressive Symptoms among Farmers. *Annals of Epidemiology.* 12: 389–94.

Swaen, G.M., VanVliet, C., Slangen, J.J. and Sturmans, F. (1992). Cancer Mortality Among Licensed Herbicide Applicators. *Scandinavian Journal of Work, Environment and Health.* 18: 201–4.

Tuliniu, H., Sigvaldason, H. and Olafsdotti, G. (1999). Breast Cancer Incidence and Familiality in Iceland During 75 years from 1921 to 1995. *Journal of Medical Genetics.* 36: 103–7.

Van Wendel de Joode, B., Wesseling, C., Kromhout, H. and Monge, P. (2001). Chronic Nervous-system Effects of Long-term Occupational Exposure to DDT. *Lancet.* 357: 1014–16.

Veldman, B.A.J., Wijn, A.M. and Knoers, N. (1998). Genetic and Environmental Risk Factors in Parkinson's Disease. *Clinical Neurology and Neurosurgery.* 100: 15–26.

Viel, J.F. and Richardson, S.T. (1993). Lymphoma, Multiple Myeloma, and Leukemia among French Farmers in Relation to Pesticide Exposure. *Social Science and Medicine.* 37: 771–77.

Wen, Y., Fu, Z., Xu, J., Tang, S., Wang, Q., Li, H., Xie G., Zhu Y., Gu Y. and Tan F. (2014). Determination of 2, 4-dichlorophenoxyacetic Acid in Air of Workplace by High-performance Liquid Chromatography. *Zhonghua Lao Dong Wei Sheng Zhi Ye Bing ZaZhi.* 32(6): 458–459.

Weiner, A. (1961). Bronchial Asthma Due to the Organic Phosphate Insecticides. *Annals of Allergy, Asthma and Immunology.* 15: 211–12.

Werner, M.H., Phuphanich, S. and Lyman, G.H. (1995). The Increasing Incidence of Malignantgliomas and Primary Central Nervous System Lymphoma in the Elderly. *Cancer.* 76: 1634–42.

Zahm, S.H., Weisenburger, D.D. and Saal, R.C. (1993). The Role of Agricultural Pesticide use in the Development of non-Hodgkin's Lymphoma in Women. *Archives of Environmental Health.* 48: 353–58.

Zheng, T., Holford, T.R. and Ma, Z. (1996). Continuing Increase in Incidence of Germ-cell Testis Cancer in Young Adults: Experience from Connecticut, USA, 1935–1992. *International Journal of Cancer.* 65: 723–29.

Zweiner, R.J. and Ginsburg, C.M. (1988). Organophosphate and Carbamate Poisoning in Infants and Children. *Pediatrics.* 81: 121–12.

4

Impact of Pesticides on Human Health

Yasmeen Khan

Pesticides are a wide range of chemicals used to get rid of a variety of undesirable organisms from the agricultural fields. These chemicals are used to protect the crops and produce more. Such chemicals are now organism specific and a number of pests are turning immune to existing pesticides. Hence, number of pesticides and quantity used for killing the pest is increasing day by day. The other dangerous and more serious phenomenon is that these chemicals persist in different components of the environment. These harmful chemicals are biologically accumulating and magnifying in the food chains, ultimately reaching the human body in higher concentrations and much faster. Consequently, number of diseases and total number of patients are increasing every moment.

Pesticides are lipophilic, bio-accumulative and non-biodegradable, and vary greatly in toxicity. Toxicity can be either acute or chronic. Exposure to pesticides can occur directly from occupational, agricultural and household use, as well as indirectly through edible items.

Persistent organic pollutants (POPs) are characterized by low water solubility and high lipid solubility, leading to bio-accumulation in fatty tissues. Pesticides are now classified as POPs. These chemicals are also known for their potential of long range transport. They are not contained within the geographic area of their release or application.

The substances generally attach themselves with such media as air, water, soil, sediments etc. and circulate.

Evidence demonstrates that certain chemicals are used in the production of many items that we use every day. Some of these organic chemicals are persistent, bio-accumulative and toxic, causing long-term harm to the health of human beings and the environment. Throughout the world, people and their environments are exposed on a daily basis to POPs. These organic compounds are resistant to photo-catalytic, biological and chemical degradation. Generally POPs are halogenated and show high solubility in fats and have very low solubility in water. These properties make POPs very potent environmental pollutants. Because of their lipophilic nature they are easily accumulated in the organism (bio-accumulation). Since the cell membrane (plasma membrane) contain 30 per cent dry weight of lipids, therefore, these pollutants easily enter the cells and being hydrophobic in nature, they cannot be excreted easily. The organism keeps accumulating these chemicals on every exposure.

POPs have contaminated almost all food items and dairy products worldwide. Fruits, vegetables, poultry and milk are all laced with high pesticide residues much above the maximum residue limits (MRL). Contamination of blood with these noxious chemicals is a matter of great concern as low (chronic) dose exposure of these chemicals in human beings may lead to health effects such as hormone disruption, reproductive abnormalities and cancer.

If the credits of pesticides include enhanced economic potential in terms of increased production of food and fibre, and amelioration of vector-borne diseases, then their debits have resulted in serious health implications to man and his environment. There is now overwhelming evidence that some of these chemicals do pose a potential risk to humans and other life forms and unwanted side-effects to the environment (Forget, 1993; Igbedioh, 1991; Jeyaratnam, 1985). No segment of the population is completely protected against the exposure to pesticides and the potentially serious health effects, though a disproportionate burden is shouldered by the people of developing countries and by high risk groups in each country (WHO, 1990). Number of deaths and chronic diseases due to pesticide poisoning is about 1 million per year worldwide (Environews Forum, 1999).

Indirect impact of pesticides on humans include consumption of food contaminated with pesticides as well as contact with pesticide residues in the air, water, soil, sediment, food materials, plants and animals. Different groups and segments of a population are exposed to pesticides in different ways and to different degrees. These are intentional (suicides and homicides) and unintentional exposures (occupational and non-occupational exposure from water, air and food). The occupational hazards in industrial settings and the ecological repercussions in the environment could be grouped as under:

(i) Operational hazards could be during manufacture and formulation of pesticides in industrial settings and their distribution and use in field conditions.

(ii) Direct toxic effects on non-target animal life such as pollinators, predators, wildlife, etc. during application of pesticides.

(iii) Post-application hazards or indirect toxic effects which involve risk to non-target animals due to toxic residues of pesticides in food or due to pollution of the ecosystem, habitat as a whole such as waterbodies or soil.

Food contamination is an ongoing public concern. The contamination can take place at any point between the fields where food is grown and food consumption. Vegetables are susceptible to insect and disease attacks easily when it is in the farm and in store. Thus, to protect it from damage, pesticides are widely used. However, some of the pesticides have residual effect and may pose serious threat to the health of the consumers (Lee and Seeneevassen, 1998). The toxicity of most pesticides and the consumption of raw fruit and vegetables reinforce the concern for contamination of these food substances over other foods (Jose et al., 2004). As a result, levels of pesticides in different food items are regulated by international and national organizations in order to protect human health (Flemming, 1981). For the protection of the public against the toxic effects of pesticides, regulatory agencies in many countries have established standards specifying the residue levels of each pesticide in various foodstuffs. At an international level, the WHO, in conjunction with the FAO, has been convening Joint FAO/WHO meetings on Pesticide Residues annually since 1961. At these meetings, the toxicological and related data are evaluated for the establishment of an acceptable daily intake or provisional tolerable daily intake (Krbic, 2007).

Table 4.1

General Pesticide Toxicity Categories and their Impact

Category	*System affected*	*Common symptoms*
Respiratory	Nose, trachea, lungs	Irritation, tight chest, coughing, choking,
Gastrointestinal	Stomach, intestines	Nausea, vomiting, diarrhoea
Renal	Kidney	Back pain, urinating more or less than usual
Neurological	Brain, spinal cord	Headache, dizziness, confusion, behaviour, depression, coma, convulsions
Haematological	Blood	Anaemia (tiredness, weakness)
Dermatological	Skin, eyes	Rashes, itching, redness, swelling
Reproductive	Ovaries, testes, foetus	Infertility, miscarriage

Use of pesticides in India began in 1948 when DDT was imported for malaria control and BHC for locust control. India started pesticide production with a manufacturing plant for DDT and benzenehexachloride (BHC) (HCH) in the year 1952. In India DDT and BHC were the two major chemicals used in agriculture and public health programmes.

Aldrin, Dieldrin, Endrin, DDT, HCH and Endosulfan had been banned in years 1996, 2003, 1990, 2011, 1997 and 2011 respectively. The half-life of Aldrin and Dieldrin is few days to 10 years, Endrin few days to 14 years, DDT is from few days to 150 years, Heptachlor few days to 14 years, HCH is from few days to 10 years and Endosulfan is from few hours to 7 years in different components (air, water, soil, animals, plants, human beings etc.) of environment.

Different types of pesticides in Indian agriculture market show that organophosphorous (50%) ranked first, followed by pyrethroids (19%), organochlorine (18%), carbamates (4%) and biopesticides (1%) (Dhaliwal and Singh, 2000). The usage of pesticides in Indian agriculture is comparatively low (0.5 kg/ha), (only 3.75% of global usage) as compared to 12.0, 7.0, 6.6 and 3.0 kg/ha in Japan, USA, Korea and Germany, respectively (Chauhan and Singhal, 2006).

Very high level of residues of organochlorine pesticides have been reported in vegetable, soil, water, blood, fat and milk samples of human beings of India (Karanth, 2002; Kumari et al., 2002; Kumar et al., 2006; Sharma and Khan, 2009; Charan, 2010; Hébert and Rondeau, 2004; PAN Europe, 2004).

Materials and Methods

Field survey was conducted for the period of one year from 2010 to 2011. Tomato was selected for the assessment of pesticide residues as it is used by people in their daily diet in various ways such as vegetable, raw in salad, in ketchup etc. and 100 samples of tomato were collected from agricultural fields. After collection, these samples were kept in polythene bags and then transported on ice to the laboratory where they were analysed, stored at 4 °C until analysis.

Extraction and Cleanup

All the solvents used in extraction and cleanup process were of HPLC grade procured from E. Merck India limited. Multi Residue Method (MRM) was followed for the extraction of organochlorine pesticides from vegetable samples and 50 gm of each vegetable was grinded and extracted twice, with 50 ml acetonitrile each time. The pooled extract was partitioned with petroleum ether (50 ml). To this extract, 5 ml of NaCl solution (2%) and 300 ml of distilled water was added. The solvent layer, after discarding aqueous layer, was washed with two 100 ml portions of distilled water. The solvent layer was then treated with 7.5 gm anhydrous sodium sulphate to remove the moisture. Extract after cleanup was evaporated till dryness and final makeup done with n-Hexane and stored in deep freezer till analysis.

Discussion and Results

The present study was undertaken to determine the concentration of different banned organochlorine pesticides' residues in vegetable (Tomato) of Sri Ganganagar district, in the state of Rajasthan. Pesticides are known to be present in vegetables due to extensive use of corresponding pesticides in interfield cultivation. The result of the study reveals that the 64 per cent of total analysed samples were contaminated with different pesticides residue. Around 45 per cent of total contaminated samples exceeded the maximum residual limit (MRL) values prescribed by the FAO/WHO. Out of hundred samples of tomato, fifty-two were found contaminated with Endosulfan and among them, seven exceeded the MRL value. Thirty-three were found contaminated with Aldrin and among them, eighteen exceeded the MRL value. Fifty were found contaminated with HCH and among them, twenty exceeded the MRL value. Ten were found contaminated with DDT. Heptachlor and Endrin was not detected in any of the samples.

The results of this study corroborate the results of a study carried in Pantnagar by Bindra (1973), Delhi by Mukherjee (2003), Kolkata by Kumar and Mukherjee (2012) which show 100 per cent contamination of vegetables. However, the detection rate in Hyderabad, Punjab and Mysore was found to be 61 per cent, 38.8 per cent and 29.5 per cent

Table 4.2

Seasonal Distribution of Organochlorine Pesticide in Tomato (in ppm)

Seasons	*ALD/Dieldrin*	*α HCH*	*β HCH*	*γ HCH*	*DDT*	*αEndo*	*βEndo*	*Endo*	*Heptachlor*	*Endrin*
Beginning season (January)	0.120 ± 0.038	ND	ND	0.1418 ± 0.098	0.0137	0.0578 ± 0.035	0.0474 ± 0.013	0.1052	ND	ND
Middle season (June-July)	0.083 ± 0.026	ND	ND	0.1352 ± 0.015	0.0126	0.0397 ± 0.004	0.0257 ± 0.009	0.0654	ND	ND
End Season (December)	0.027 ± 0.008	ND	ND	0.1268 ± 0.060	0.0063	0.0325 ± 0.005	0.0191 ± 0.003	0.0516	ND	ND

ND = Not Detected

Values are ± SD of 30 observations

Table 4.3

Total Organochlorine Pesticide Residues in Tomato (in ppm)

Name of pesticide detected	*Vegetables (No. of samples studied)*	*Total no. of samples contaminated with different pesticides*	*Acute toxicity*	*MRL In ppm WHO/FAO*	*% of samples contaminated with pesticides*	*No. of samples exceeded MRL in ppm*
Endosulfan		52	High	0.5	52%	7
Aldrin		33	High	0.1	33%	18
Dieldrin	Tomato	ND	High	0.1	ND	Nil
HCH (Lindane)	(100)	50	Moderate	0.01	50%	20
DDT		10	Moderate	Not Mentioned	1%	NA

respectively. A very high per cent of vegetables from Hyderabad, Ludhiana and Mysore have been reported to be positive for pesticides (70.7%).

A comparison of results of this study was made with other reports to assess the status of OCP contamination in vegetables (tomato) from Sri Ganganagar. The study shows that ÓHCH, ÓDDT and Ó Endosulphan were 141.8 ppb, 13.7 ppb and 105.2 ppb respectively in tomatoes. Levels of OCPs in this study was much higher than tomatoes from Meerut, Muzaffarnagar and Ghaziabad districts of Uttar Pradesh, India (IPEN, 2006) which have 3.42–11.18 ppb, 23.9 ppb and ND respectively, Faisalabad, Pakistan, (Randhawa et al., 2007) which have ND, ND and 556 ppb respectively, Kumasi, Ghana, (Essumang et al., 2008) which have ND, ND and 260 ppb respectively, Uttarakhand, India, (Chauhan et al., 2012) which have ND, ND and 90.2 ppb respectively, Kothapally, A.P, India, (Ranga Rao et al., 2008) which have ND, ND and 32.0 ppb respectively, Ajmer, India (Charan et al., 2010) which have ND, ND and 0.28 ppb respectively and Kolkata, (Kumar and Mukherjee, 2012) which have 1.77, 1.61 and 0.11 ppb respectively.

References

Bindra, O.S. (1973). The Magnitude of Pesticidal Pollution in India. In: Bindra O.S., Kalra, R.L. (eds). *Progress and Problems in Pesticide-residus Analysis.* Ludhiana: Punjab Agricultural University and Indian Council of Agricultural Research. 41–51.

Charan, P.D., Ali, S.F., Kachhawa, Y. and Sharma, K.C. (2010). Monitoring of Pesticide Residues in Farmgate Vegetables of Central Aravalli Region of Western India. *American-Eurasian J. Agric. & Environ. Sci.* 7 (3): 255–258.

Chauhan, R.S. and Singhal, L. (2006). Harmful Effects of Pesticides and Their Control Through Cowpathy. *Int. J. Cow. Sci.* 2: 61–70.

Chauhan, S.S., Negi, S., Singh, N., Bhatia, G. and Srivastava, A. (2012). Monitoring of Pesticides Residues in Farmgate Vegetables of Uttarakhand, India. *Wudpecker Journal of Agricultural Research.* 1(7): 250–256.

Dhaliwal, G.S. and Singh, B. (2000). *Pesticides and Environment.* Commonwealth Publishers, New Delhi.

Environews Forum. (1999). Killer Environment. *Environ Health Perspect.* 107: A62.

Essumang, D.K., Dodoo, D.K., Adokoh, C.K. and Fumador, E.A. (2008). Analysis of some Pesticide Residues in Tomatoes in Ghana. *Human and Ecological Risk Assessment.* 14: 796–806.

Flemming, H.F. (1981). Analysis of Residues of Seven Pesticides in Some Fruits and Vegetables by Means of High Pressure Liquid Chromatography. *J.F. Bergm. Vertag.* 173: 95–98.

Forget, G. (1993). Balancing the Need for Pesticides with the Risk to Human Health. In: Forget G., Goodman T., de Villiers A. (eds). *Impact of Pesticide Use on Health in Developing Countries.* IDRC, Ottawa: 2.

Hébert, S. and Rondeau, B. (2004). Saint Laurent Vision 2000 Actions Plan. Phase III: The Water Quality of Lake Saint-Pierre and its Tributaries. http://www.slv2000.qc.ca/plan_action/phase3/biodiversite/suivi_ecosysteme/atelier_20041203/presentations/SH_qual_eaux_a.htm.

Igbedioh, S.O. (1991). Effects of Agricultural Pesticides on Humans, Animals and Higher Plants in Developing Countries. *Arch Environ Health.* 46: 218.

International POPs Elimination Network (2006). Establishing the Prevalence of POPs Pesticide Residues in Water, Soil and Vegetable Samples and Creating Awareness about their ill-effects. (http://www.ipen.org).

Jeyaratnam, J. (1985). Health Problems of Pesticide Usage in the Third World. *British Medical Journal.* 42: 505.

Jose, L.T., Consuelo, S.B., Beatriz, A. and Lorena, G. (2004). Analysis of Pesticide Residues in Juice and Beverage. Crit. Rev. in anal.chem. *Journal of Plant Protection.* 26(2): 145–148.

Karanth, N.G.K. (2002). Challenges of Limiting Pesticide Residues in Fresh Vegetables: The Indian Experience. In: Hanak E., Boutrif E., Fabre P., Pineiro M., (Scientific Editors). *Proceedings of the International Workshop,* CIRAD-FAO, 11–13, Montpellier, France, CIRADFAO.CIRAD CD-ROM, Montpellier, France.

Krbic, B.S. (2007). Organochlorine and Organophosphate Pesticide Residues in Wheat Varieties from Serbia. *F. Addi. and Contam.* 24: 695–703.

Kumar, A., Baroth, A., Soni, I., Bhatanagar, P., and John, P.J. (2006). Organochlorine Pesticide Residues in Milk and Blood of Women from Anupgarh, Rajasthan, India. *Environmental Monitoring and Assessment.*116(1–3): 1–7.

Kumar, B. and Mukherjee, D.P. (2012). Organochlorine Residues in Vegetables. *International Journal of Vegetable Science.* 18: 121–136.

Kumari, B., Madan, V.K., Kumar, R. and Kathpal, T.S. (2002). Monitoring of Seasonal Vegetables for Pesticide Residues. *Environmental Monitoring and Assessment.* 74(3): 263–270.

Lee, F.C.L.H. and Seeneevassen, S. (1998). Monitoring Insecticide Residues in Vegetables and Fruits at the Market Level. 95–102. In FARC, (eds). Annual Meeting of Agricultural Scientists. Food and Agricultural Research Council. Readuit, Mauritius.

Mukherjee, I. (2003). Pesticides Residues in Vegetables in and Around Delhi. *Journal of Environment Monitoring Assessment.* 86(3): 265–71.

PAN Europe. (2004). Pesticide Action Network Europe: Pesticides in food – what's the problem? Briefing no. 3, September 2004, Facilitated by PAN Germany and PAN UK. http://www.pan-europe.info/publications/index.shtm.

Randhawa, M.A., Anjum, F.M., Ase, M.R., Bhutt, M.S., Ahmed, A. and Randhawa, M. S. (2007). Removal of Endosulphan Residues from Vegetables by Household Processing. *J. Sci. Ind. Res.* 66: 849–852.

Ranga Rao, G.V., Sahrawat, K.L., Rao, Ch. S., Das, B., Reddy, K.K., Bharath, B.S., Rao, V.R., Murthy, K.V.S. and Wani, S.P. (2009). Insecticide Residue in Vegetable Crops Grown in Kothapalli Watershed, Andhra Pradesh, India: A Case Study. *Indian Journal of Dryland Agriculture Research And Development.* 24(2): 21–27.

Sharma, A. and Khan, T.I. (2009). Status of Organochlorine Insecticides in Groundwater: A Case Study of Agriculture Intensive Areas of Jaipur. *Journal The Ecoscan.* 3 (1 and 2): 83–86.

WHO (1990). Public Health Impact of Pesticides used in Agriculture. World Health Organization in Collaboration with the United Nations Environment Programme, Geneva.

Section–II
Fluoride and Health

5

Fluoride and Health Hazards

Nisha Jain and S.C. Joshi

Water is an essential valuable natural resource for sustaining life and environment, which we have always thought to be available in abundance and a free gift of nature. However, chemical composition of surface water or subsurface, geothermal or non-thermal, is one of the prime factors on which the suitability of the water for domestic, industrial or agriculture purpose depends. In India, 30 per cent of urban and 90 per cent of rural population is dependent on an untreated water source, of which 80 per cent is groundwater. This groundwater when consumed for drinking purposes has various toxic effects. High groundwater fluoride concentrations associated with igneous and metamorphic rocks such as granites and gneisses have been reported. As a result, people of 19 states in India are drinking fluoride contaminated water above the WHO maximum allowed concentration of 1.5 mg/l. Fluoride has beneficial effects on teeth at low concentrations in drinking water, but excessive exposure to fluoride in drinking water or in combination with exposure to fluoride from other sources, can give rise to a number of adverse effects. These range from mild dental fluorosis to crippling skeletal fluorosis as the level and period of exposure increases. Apart from these, there are reports that excess fluoride consumption promotes cancer, hip fracture, leads to still-births or birth defects and has detrimental neurological effects.

International Status

Chronic fluoride intoxication (fluorosis) is a worldwide health problem wherever drinking water contains high amount of fluoride. The primary manifestation of fluorosis is mottling of teeth and oesteosclerosis and secondary neurological complication (para and quadriplegia) have been observed both in humans and animals. High concentration of fluoride has been detected in underground water in many countries, e.g. Kenya, Nigeria, South Africa, Tanzania, USA, China, India, Japan, Korea and Australia. Besides, 33 other countries have been reported to have high fluoride level in the groundwater.

National Status

India lies in a geographical fluoride belt which extends from Turkey to China and Japan through Iraq, Iran and Afghanistan. Of the 85 million tonnes of fluoride deposits found on the earth's crust, nearly 12 million tonnes are in India. In India, fluorosis is endemic in many states because groundwater in more than one-third of an Indian district is not fit for drinking. Fluoride contaminated groundwater is creating health problems in India. Nearly 90 million people, including 6 million children, in the country in 200 districts in 15 states are affected with dental, skeletal and non-skeletal fluorosis. It occurs in children during the developmental stages when the teeth are exposed to fluoride. The extent of fluoride contamination in groundwater varies from 1.0-48 mg/l. The fluoride affected states are Andhra Pradesh, Bihar, Delhi, Gujarat, Haryana, Jammu and Kashmir, Karnataka, Kerala, Madhya Pradesh, Maharashtra, Orissa (now Odisha), Punjab, Rajasthan, Tamil Nadu and Uttar Pradesh (Figure 5.1).

State of Rajasthan

In Rajasthan, groundwater is characterized by comparatively high concentration of fluoride. Around 31 out of 33 districts in Rajasthan are presently consuming water contaminated with fluoride. As many as 9,741 villages, 6,819 habitations has fluoride levels of more than 1.5 mg/l. Rajasthan has an area of 3.42 lakh sq. km, the largest state of country having 10.41 per cent of the country's area and 5.5 per cent of the nation's population but has low water resources (just 1% of the country's resources). The state has extreme climatic and geographical conditions. The Aravalli range of hills divides the state into two parts. The north-west region has scanty and uncertain rainfall while the south-east region has fairly good rainfall. Thus, the state has water problem of quality and quantity both. In most parts of the state,

Figure 5.1

Fluoride Affected States in India

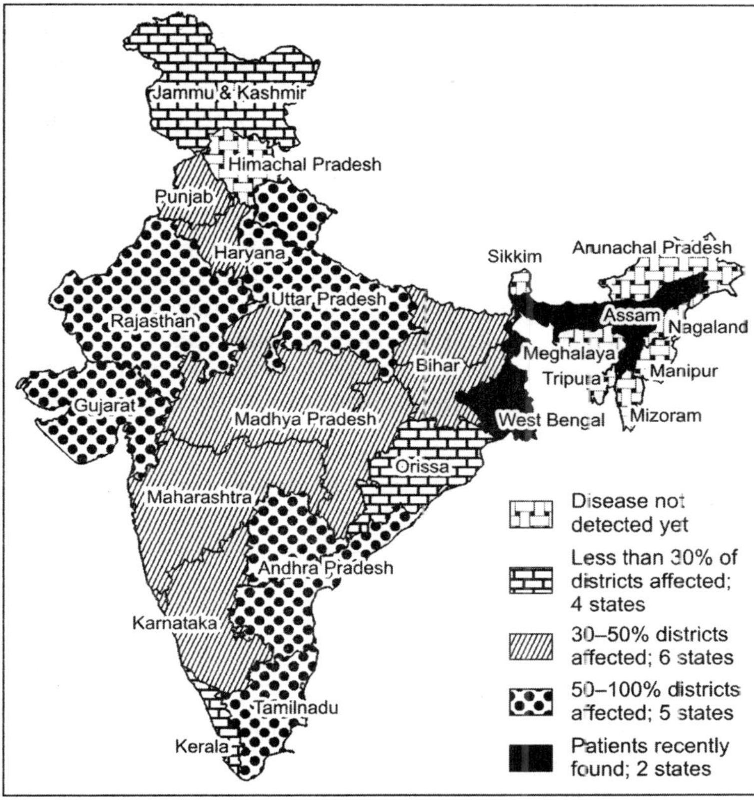

groundwater is either saline or has excess fluoride. Groundwater is the major source of drinking water and over 91 per cent of the drinking water demand is met by groundwater.

Sources of Fluoride in Water

Fluorine always occurs in combination with minerals as fluoride. It has high reactivity and represents about 0.06 to 0.09 per cent of the earth crust (WHO, 1996). The presence of fluorine in groundwater is mainly a natural phenomenon, and majorly influenced by local and regional geological conditions, as the fluoride minerals are nearly insoluble in water. Hence, fluorine is present in groundwater only when the conditions favour their solution. The main source of fluorine in groundwater is basically from the mafic minerals. These minerals are commonly associated with the country rocks through which the groundwater percolates under variable temperature conditions. Besides these minerals, alkali rocks, hydrothermal solutions may also contribute to higher concentration of fluoride in groundwater. Robinson and Edington (1946) reported that the main source of fluorine in ordinary soil consists of clay minerals. The weathering and leaching process occurs mainly by moving and percolating.

Fluoride and Human Health

Fluoride contamination is a major health hazard in many parts of the world. Fluoride is considered beneficial to human health if taken in limited quantity (0.5 to 1.5 mg/l). Fluoride prevents tooth decay by enhancing the demineralization of enamel that is under attack, as well as inhibiting the production of acid by decay causing bacteria in dental plaque. Fluoride is also a normal constituent of the enamel itself.

Table 5.1

Effects of Fluoride in Drinking Water

S. No.	*Fluoride in drinking water (mg/l) effect*
1.	0.002 mg/l in air Injury to vegetation
2.	1 mg/l in water Dental caries reduction
3.	2 mg/l or more in water Mottled enamel
4.	3.1 to 6.0 mg/l in water Osteoporosis
5.	8 mg/l in water Osteoporosis
6.	20–80 mg/day or more in water or air crippling skeletal fluorosis
7.	50 mg/l in food or water Thyroid change
8.	100 mg/l in food or water Growth retardation
9.	More than 125 mg/l in food or water Kidney damage
10.	2.5–5.0 gm in actual dose Death

It is also known to cause dental, skeletal fluorosis, osteosclerosis, thyroid, kidney changes and cardiovascular, gastrointestinal, endocrine, neurological, reproductive, developmental, molecular level, immunity effects, etc.

Effect on Dental Enamel

Dental fluorosis is a condition that results from the intake of excess levels of fluoride during the period of tooth development, usually from birth to approximately 6–8 years of age. It has been termed a hyper-plasia or hypo-mineralization of dental enamel and dentine and is associated with the excessive incorporation of fluoride into these structures. The severity of this condition, generally characterized as ranging from very mild to severe, is related to the extent of fluoride exposure during the period of tooth development. Mild dental fluorosis is usually typified by the appearance of small white areas in the enamel; individuals with severe dental fluorosis have teeth that are stained and pitted ('mottled') in appearance. Experimental animal studies suggest that this hypo-mineralization results from fluoride disturbance of the process of enamel maturation (Richards et al., 1986). In India, Viswanathan (1951) first reported a disease similar to mottled enamel, which was prevalent in human beings in Madras presidency. Mahajan (1934) reported a similar disease in cattle in certain parts of old Hyderabad state. However (Shortt et al., 1937), was the first to identify the disease as fluorosis. Subsequent to these findings, cases of fluorosis were reported from several other parts of the country.

Table 5.2

Dean's Index

Classification	*Criteria-description of Enamel*
Normal	Smooth, glossy, pale creamy-white translucent surface
Questionable	A few white flecks or white spots
Very Mild	Small opaque, paper white areas covering less than 25 per cent of the tooth surface
Mild	Opaque white areas covering less than 50 per cent of the tooth surface
Moderate	All tooth surfaces affected; marked wear on biting surfaces; brown stain may be present
Severe	All tooth surfaces affected; discrete or confluent pitting; brown stain present

Figure 5.2

Dental Fluorosis in School going Girls of Jaipur District

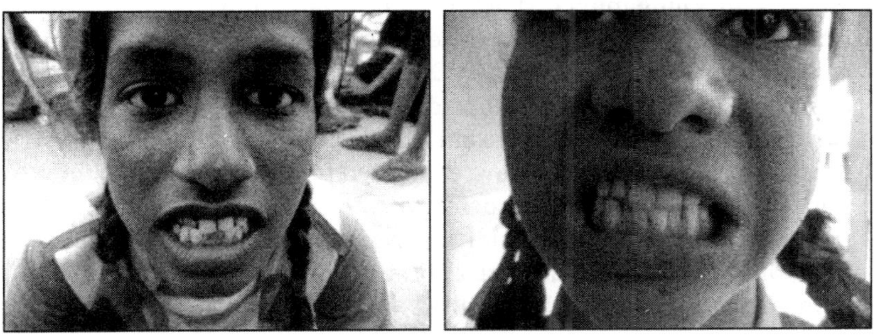

Osteoporosis: Fluoride above 4 mg/l in drinking water may cause a condition of dense and brittle bones known as osteoporosis. It affects millions of people worldwide (Hussain et al., 2004) and is responsible for as many as 75 per cent of all fractures in people over the age of 45. Costly and disabling fractures of spine, hip, wrist and other bones can be preceded by years of undetected bone loss. It is found that as many as 20 per cent of those who suffer from osteoporosis related hip fractures die within 6 months. Women are at four times greater risk of developing osteoporosis than males (Bezerra et al., 2003).

Skeleton Fluorosis: The chronic toxic effects of fluoride on the skeletal system have been described from certain geographical regions of the world where drinking water contains excessive quantities of natural fluoride. This form of chronic intoxication was first described in India from the state of Madras as early as 1937 (Shortt et al., 1937). Subsequently cases of endemic fluorosis have been reported from other parts of India, particularly from Punjab and sporadically from other parts of the world, notably Ceylon, China, Japan, Saudi Arabia (El Tannir, 1959), USA (Leone et al., 1954; Zipkin et al., 1958) Canada and Europe. Besides endemic fluorosis, chronic toxic effects of fluoride on the skeletal system have also been observed in relation to industrial exposure to fluorides such as cryolite and in fact it is the pioneering studies of Roholm (1937) that paved the way for further contributions on the subject. At higher levels of ingestion (2 to 8 mg daily) signs of fluorosis appear in teeth as mineralization, certain other factors (climatic conditions, malnutrition, age, storage, other constituents of water and possibly individual variations in absorption) may be involved. Under such conditions and over a number of years, skeletal fluorosis may arise characterized by an increased density of bone and demonstrated in adults radio graphically (Yildiz et al., 2003). A correlation between average water fluoride concentration and prevalence of skeletal fluorosis (assessed by X-ray) was found among adults in 15 villages in Dungapur district in Rajasthan, India. The prevalence ranged from 4.4 per cent at a water fluoride level of 1.4 mg/l to 63 per cent at the level of 6.0 mg/l. Crippling fluorosis was consistently observed in villages with fluoride concentrations of >3 mg/l. In a survey carried out by (Hussain et al., 2004) in Bhilwara district, Rajasthan, 825 individual were examined for fluorosis due to intake of fluoride above 5.0 mg/l in drinking water. In the skeletal fluorosis positive individuals maximum individual (194 individuals, 23.52%) have Grade I skeletal fluorosis that is characterized by bone and joint pain. Only 4 individuals (0.48%) have Grade III type of skeletal fluorosis characterised by bone and joint pain, stiffness and rigidity of dorso-lumbar spine and restricted movement at spine and joints which are general symptom including deformities of spine and limbs, knock knees, crippled or bed ridden state, kyphosis, invalidism, genuvarum and genu-valgum. Prevalence and severity of skeletal fluorosis was found to increase with rise in fluoride concentration.

Deformities and Crippling Fluorosis: The advanced stage of fluoride intoxication results from the continuous exposure of an individual to 20–80 mg of fluoride ion daily over a period of 10–20 years. Such heavy exposure is associated with a level of at least 10 mg/l in the drinking water supply. The crippling fluorosis is seen in such numbers in endemic areas of Rajasthan, Punjab, Orissa (Odisha) and southern India. The crippling deformities are due partly to mechanical factors and partly to the immobilization necessitated by pain and paraplegia. The commonest deformities are kyphosis, flexion deformity of the hips, flexion deformity of the knees and fixation of the chest in the position of inspiration due to calcification of cartilages. The quadriplegic patient is bent with kyphosis and with restricted movements of his spine, contractures of hips and knees (Kaminsky et al., 1990).

Cardiovascular Effects: The cardiovascular effects of fluoride have been attributed to hypocalcaemia and hyperaemia caused by high fluoride levels. Fluoride can bind with serum calcium if the dose is sufficient and cause hypocalcaemia. Calcium is necessary for the functional integrity of the voluntary and autonomic nervous systems. Hypocalcaemia can cause tetany, decreased myocardial contractility, and possibly cardiovascular collapse (Bayless and Tinanoff, 1985). Hyperaemia has been suggested as the cause of the repeated episodes of ventricular fibrillation and eventual death that are often encountered in cases of fluoride poisoning (Baltazar et al., 1980).

Gastrointestinal Effects: The primary gastrointestinal effects following both acute and chronic oral exposure to fluoride consist of nausea, vomiting, and gastric pain. The irritation of the gastric mucosa is attributed to fluoride (as sodium fluoride) forming hydrofluoric acid in the acidic environment of the stomach (Hoffman et al., 1980; Waldbott, 1981). The uncharged hydrogen fluoride molecule can then penetrate cell membranes and enter the neutral environment of the cytoplasm. A study assessed the prevalence and severity of gastrointestinal disturbance in an area of endemic skeletal and dental fluorosis in India. The highest prevalence (52.4%) of non-ulcer dyspeptic symptoms was found among 288 individuals (69 families) living in a village where the mean fluoride concentration in the 36 separate water sources was 3.2 ppm (range 0.25 to 8.0 ppm). Eleven of these water sources were defined by the authors as safe (i.e. with fluoride levels of 1.0 ppm or less). The authors noted that in patients who reverted to safe water, dyspeptic symptoms and complaints disappeared within 2–3 weeks.

Endocrine Effects: In the endocrine system where the intermediary metabolism and synthesis of highly sensitive hormones involves enzymatic action, it is expected that interferences with the mechanism by chemical agents would produce early and pronounced clinical effects. Considerable attention has consequently been given in recent years to the behaviour of fluoride in hormone chemistry and to the possible clinical disturbances of endocrine function, particularly the thyroid gland. Significant increases in serum thyroxin levels were observed in residents of north Gujarat, India with high levels of fluoride in the drinking water (range of 1.0–6.53 mg/l; mean of 2.70 mg/l). No significant changes in serum tri-iodo thyronine or thyroid stimulating hormone levels were found. Increases in serum epinephrine and norepinephrine levels were also observed. It is unclear if nutritional deficiencies played a contributing role to the observed endocrine effects.

Immunological and Lymph Reticular Effects: A request to the American Academy of Allergy was made by the US Public Health Service for an evaluation of suspected allergic reactions to fluoride as used in the fluoridation of community water supplies (Austen et al., 1971). The response to this request included a review of clinical reports and an opinion as to whether these reports constituted valid evidence of a hypersensitivity reaction to fluoride exposure of types I, II, III, or IV (Austen et al., 1971), which are, respectively, anaphylactic or reaginic, cytotoxic, toxic complex, and delayed-type reactivity. The Academy reviewed the wide variety of symptoms presented (vomiting, abdominal pain, headaches, scotomata (blind, or partially blind areas in the visual field), personality change, muscular weakness, painful numbness in extremities, joint pain, migraine headaches, dryness in the mouth, oral ulcers, convulsions, mental deterioration, colitis, pelvic haemorrhages, urticaria nasal congestion, skin rashes, epigastria distress, and hematemesis) and concluded that none of these symptoms were likely to be immunologically mediated reactions of types I–IV. No studies were located that investigated alterations in immune response following fluoride exposure in humans. In a study with rabbits administered 4.5 mg fluoride/kg/day as sodium fluoride for 18 months, decreased antibody titters were observed (Jain and Susheela, 1987). These results were observed after 6 months of treatment; the authors hypothesized that a threshold level is reached at which time the immune system is impaired.

Neurological Effects: The neurological manifestations have been exclusively reported from India. Credit for the earliest description of neurological complications in fluorosis must be given to Shortt et al. (1937), who reported ten such cases from the Nellore district of Madras (Tamil Nadu). A few sporadic cases have also been described from other parts of India (Murthi et al., 1953). Fluoride has been shown to interfere with glycolysis. Because the central nervous system relies heavily on this energy source, hypotheses have been advanced as to a mechanism for fluoride effects on the central nervous system. Although effects on glycolytic enzymes could explain the neuromuscular symptoms seen frequently in cases of fluoride poisoning (e.g. tetany, paresthesia, paresis, convulsions), studies tend to indicate that hypocalcaemia caused by fluoride binding of calcium causes these symptoms. The decreases in intelligence were reported in children living in areas of China with high levels of fluoride in the drinking water, as compared to matched groups of children living in areas with low levels of fluoride in the drinking water (Li et al., 1995 and Lu et al., 2000), but these studies are weak in as much as they do not address important confounding factors.

Reproductive Effects: There are limited data on the potential of fluoride to induce reproductive effects in humans following oral exposure. A met analysis found a statistically significant association between decreasing total fertility rate and increasing fluoride levels in municipal drinking water. A study found significantly decreased serum testosterone levels in 30 men diagnosed with skeletal fluorosis and in 16 men related to men with fluorosis and living in the same house as the patient (Susheela and Jethanandani, 1996). The mean drinking water fluoride levels were 3.9 ppm (approximately 0.11 mg fluoride/kg/day), 4.5 ppm (0.13 mg fluoride/kg/day), and 0.5 ppm (0.014 mg fluoride/kg/day) in the patients with skeletal fluorosis, related men, and a control group of 26 men living in areas with low endemic fluoride levels. No correlations between serum testosterone and urinary fluoride levels or serum testosterone and serum fluoride levels were found. One limitation of this study was that the control men were younger (28.7 years) than those with skeletal fluorosis (39.6 years) and the related men (38.7 years). In addition, the groups are small and potentially confounding factors are not well addressed.

Developmental Effects: Fluoride crosses the placenta in limited amounts and is found in foetal and placental tissue. The available human data suggest that fluoride has the potential to be developmentally toxic at doses associated with moderate to severe fluorosis. The human and animal data suggest that the developing foetus is not a sensitive target of fluoride toxicity. Analysis of birth certificates and hospital records for over 2,00,000 babies born in an area with fluoridated water and over 10,00,000 babies born in a low fluoride area, found no difference in the incidence of birth defects attributable to fluoride (Erickson et al., 1976). Exposure to high levels of fluoride has been described together with an increased incidence of spinal bifida (Gupta et al., 1995). The occurrence of spinal bifida was examined in a group of 50 children aged 5–12 years living in an area of India with high levels of fluoride in the drinking water (4.5–8.5 ppm) and manifesting into either clinical (bone and joint pain, stiffness and rigidity), dental or skeletal fluorosis. An age- and weight-matched group of children living in areas with lower fluoride levels (1.5 ppm) served as a control group. Spinal bifida was found in 22 (44%) of the children in the high fluoride area and in six (12%) children in the control group. This study however, did not examine the possible role of potentially important nutrients such as folic acid, and had other study design flaws.

Fluoride and IQ: A total of 56 studies have investigated the relationship between fluoride and human intelligence, and a total of 36 studies have investigated the relationship between fluoride and learning/memory in animals. Of these investigations, 49 of the 56 human studies have found that elevated fluoride exposure is associated with reduced IQ, while 34 of the 36 animal studies have found that fluoride exposure impairs the learning and/or memory capacity of animals. The human studies, which are based on IQ examinations of over 12,000 children, provide compelling evidence that fluoride exposure during the early years of life can damage a child's developing brain.

Effect at Molecular Level: The acceleration of the ageing process by fluoride occurs at the bio-chemical level through enzyme inhabitation, collagen breakdown, genetic damage or disruption of the immune system. Fluoride damage enzymes and results in a wide range of chronic disease. Fluoride, as low as 1 mg/l results in breakdown of collagen and the most abundant of the body protein at 30 per cent. It leads to irregular formation of collagen which serves as a major structural component of skin, ligaments, tendons, muscles; fluoride interferes with a number of physiological processes. Most evidences indicates that it acts on the DNA repair enzyme system. It may also interfere with DNA synthesis. If unprepared, DNA damage occurs in a cell producing sperm or egg, it will be replicated in every cell of the offspring body and leads to birth defects. Irreparable damage of a segment of DNA is responsible for control of cell growth and may cause tumours or cancer.

Effect on Immune System: Fluoride interacts with the bonds of protein molecular required to maintain the normal shape of proteins. Fluoride affects the immune system:

(i) Damage to the immune system by inhibiting the migration rate of white blood cells to infected means, (ii) Interferes with phagocytosis (destruction of bacteria and other foreign agents by white blood cells, or (iii) Induces the release of super oxide free radicals in resting white blood cells (Shailaja and Johnson, 2007). The fluoride induced interference leads to an increased and more prolonged exposure of the body to foreign materials and releases free radicals damaging the body.

Fluoride as Carcinogen: Fluoride was found to be an equivocal carcinogen by the National Cancer Institute. Toxicological effects due to excess of fluoride is 6.9 times and cause bone cancer in young males. Several epidemiological studies are available on the possible association between fluoride in drinking water and cancer rates among the population Indian Association for Cancer Research (IACR) evaluated these studies on fluoride and carcinogenicity in humans (IARC, 1982). Numerous epidemiological studies have examined the issue of a connection between fluoridated water and cancer. The weight of evidence indicates that no such connection exists. However, all of the investigations were ecologic studies, and this sensitivity limit of even the most sensitive analysis in these studies appears to be a 10–20 per cent increase. Since any carcinogenic effect of fluoride at the levels found in water supplies would probably be below this level of sensitivity, a National Toxicology Programme (NTP) cancer bioassay was conducted to assess the effect of fluoride on cancer incidence in animals (NTP, 1990). The NTP study found equivocal evidence of a fluoride-related increase in osteo-sarcomas in male rats, and no evidence of any fluoride-related neoplasm in female rats or male or female mice. A study sponsored by Proctor and Gamble found no evidence of fluoride carcinogenicity in either male or female rats.

Fluoride Mitigations

The challenges in the fighting fluoride menace can be met only through a well-planned strategy carried out diligently by the government and other stakeholders in close coordination with each other. The list of measures recommended is by no means exhaustive or complete but are enumerated below:

1. *Creating Public Awareness:* The government needs to be at forefront to create public awareness campaign on the lines of the ones we have on TB/leprosy and HIV/AIDS. The media - visual media, in particular – also has an important role in the spread of awareness about the fluoride problem.
2. *Regular Monitoring of Water Quality:* Water quality monitoring can no more remain a one-off or occasional activity. It has to be a rigorous, year-round exercise involving the testing of drinking water on all 15 parameters recommended by the centre.
3. *Sharing of Water Quality Information with the Public:* The water quality information collected by the government agencies during the process of monitoring is useless unless it is shared with the public at large and the population that is going to be affected, in particular.
4. *Providing Alternative Sources of Safe Drinking Water:* The moment excess fluoride levels have been detected in a drinking water source, it needs to be closed to public – but not before alternative sources of safe drinking water have been provided, as the government is doing now. Money should not be a consideration in the matter of providing safe drinking water to the people.
5. *Networking Between Health and Public Engineering Professionals:* Engineers specializing in public health need to know about the characteristics of fluorosis and the defluoridation technology. Networking between health and public health engineering departments will go a long way in ensuring early detection and timely remedial measures.
6. *Inter-Linking District Headquarters Hospitals:* District headquarters need to be inter-linked to provide timely referral services to victims of fluorosis.
7. *Orientation of Doctors:* There is a need for a proper, well-planned orientation programme to educate doctors about the latest findings in the field of flurosis control.
8. *Sustainable Operation and Maintenance of Drinking Water Sources:* One of the major reasons for excess fluoride in drinking

water is over withdrawal of groundwater. The community has to be involved in the management of local water resources.

9. *Issuance of Handicapped Certificate to Fluoride Victims:* The Government should issue handicapped certificate to these severely affected by fluorosis.
10. *Epidemiological Survey:* An exhaustive, statewide, house-to-house survey needs to be carried out to assess the actual magnitude of the fluoride problem. During the survey, all the three forms of fluorisis, viz. dental, skeletal and non-skeletal should be identified.
11. *Blood (Serum) and Urine Testing:* To confirm fluorosis, it is necessary to test the fluoride content in blood (serum), urine and drinking water of a patient.
12. *Nutritional Intervention:* There also is a need for nutritional intervention. The importance of consuming food rich in calcium, Vitamins C, E and anti-oxidants on a daily basis should be emphasized.
13. *Providing Cheap Defluoridation Technology to People:* Cost effective, easy to handle/operate defluoridation technology should be made available to the rural population, who bear the brunt of fluoride contamination.
14. *Water Harvesting as Alternative Water Source:* Rainwater harvesting is the best possible option to make fluoride-free water available to both humans and domestic animals.

Recent Research in Defluoridation

- Treatment of high fluoride concentration water by MgAl-CO_3 layered double hydroxides.
- Adsorption of fluoride from water by amorphous alumina supported on carbon nano tubes
- With the use of various oxide ores such as refractory grade bauxite, feed bauxite, manganese ore and hydrated oxides of manganese ores (WAD).
- With the use of natural materials such as red soil, charcoal, brick, fly-ash and serpentine.
- Red soil has good fluoride removal capacity followed by brick, fly-ash, serpentine and charcoal.

References

Austen, K.F., Dworetzky, M. and Farr, R.S. (1971). A Statement on the Question of Allergy to Fluoride. *Journal of Allergy*. 47: 347–348.

Baltazar, R.F., Mower, M.M. and Funk, M. (1980). Acute Fluoride Poisoning Leading to Fatal Hyperkalemia. *Chest*. 78(4): 660–663.

Bayless, J.M. and Tinanoff, N. (1985). Diagnosis and Treatment of Acute Fluoride Toxicity. *American Journal of Dental Association*. 110: 209–211.

Bezerra, de Menezes, L.M., Volpato, M.C. and Rosalen, P.L. and Cury, J.A. (2003). Bone as a Biomarker for Acute Fluoride Toxicity. *International Forensic Science*. 137(2–3): 209–214.

El Tannir, M.D. (1959). Mottling of enamel in Mecca and the Arabian. Peninsula. American Journal of Public Health. 49: 45–52.

Erickson, J.D., Oakley, G.P. Jr. and Flynt, J.W. Jr. (1976). Water Fluoridation and Congenital Malformations: No Association. *American Journal of Dental Association*. 93: 981–984.

Gupta, S.K., Gupta, R.C. and Seth, A.K. (1995). Increased Incidence of Spina Bifida Occulta in Fluorosis Prone Areas. *Acta. Paediatr.Jpn.*, 37(4): 503–506.

Hoffman, R., Mann, J. and Calderone, J. (1980). Acute Fluoride Poisoning in a New Mexico Elementary School. *Pediatrics*. 65: 897–900.

Hussain, J., Sharma, K.C. and Hussain, I. (2004). Fluoride In Drinking Water And Its Ill Affect on Human Health: A Review. *Journal of Tissue Research*. 4(2): 263–273.

IARC. (1982). IARC Monographs on the Evaluation of Carcinogenic Risk of Chemicals to Humans. Lyon, France, World Health Organization, International Agency for Research on Cancer. 27: 237–303.

Jain, S.K. and Susheela, A.K. (1987). Effect of Sodium Fluoride on Antibody Formation in Rabbits. *Environmental Research*. 44: 117–125.

Kaminsky, L., Mahoney, M., Leach, J., Melius, J. and Miller, M. (1990). Crit. Rev. *Oral Medical Biology*. 1: 261–281.

Leone, N.C., Shimkin, M.B., Arnold, F.A., Stevenson, C.A., Zimmerman, E.R and Lieberman, J.E. (1954). Medical Aspects of Excessive Fluoride. *Publication Health Report (Wash)*. 69: 925–936.

Li X, S., Zhi, J.L., Gao, R.O. (1995). Effect of Fluoride Exposure on Intelligence in Children. *Fluoride*. 28(4): 189–192.

Lu, Y., Sun, Z.R. and Wu, L.N. (2000). Effect of High-fluoride Water on Intelligence in Children. *Fluoride*. 33(2): 74–78.

Mahajan, J. (1934). Annual Report. VIO Hyderabad State 3, Indian Council of Agricultural Research, New Delhi.

Murthi, G.V.S., Narayana Rao, D. and Venkateswarlu, P. (1953). Studies of Endemic Fluorosis in Nalgoda District. *Indian Journal of Medical Association*. 22: 396–399.

NTP (1990). National Toxicology Program. NTP TR 393, NIH Publication No. 90. 2848.

Richards, A., Kragstrup, J., Josephsen, K. and Fejerskov, O. (1986). Dental Fluorosis Developed in Post-secretory Enamel. *Journal of Dental Residence*. 65: 1406–1409.

Robinson, W.D. and Edington, G. (1946). Fluorine in Soils. *Soil Science*. 61: 341–353.

Roholm, K. (1937). Fluorine Intoxication: A Clinical Hygienic Study with a Review of the Experimental Investigations, Copenhagen H.K. Lewis & Co. Ltd. 136, London.

Shailaja, K. and Jhonson, M.E.C. (2007). Fluorides in Groundwater and Its Impact on Health. *Journal of Environmental Biology*. 28: 331–332.

Shortt, H.E., Pandit, C.G. and Raghavachari, T.N.S. (1937). Endemic Fluorosis in the Nellore District of South India. *Indian Medical. Gaz*. 72: 398.

Susheela, A.K. and Jethanandani P. (1996). Circulating Testosterone Levels in Skeletal Fluorosis Patients . *Clinically Toxicology*. 34(2): 183–189.

Viswanathan, G.R. (1951). Annual Report Madras. Indian Council of Agricultural Research, New Delhi, 1935. (Quoted from Indian Institute of Science, 33A, I. 1951).

Waldbott, G.L. (1981). Mass Intoxication From Accidental Over Fluoridation of Drinking Water. *Clinically Toxicology*. 18: 531–541.

WHO (1996). *Guideline for Drinking Water Quality*, Second Edition, Vol. 2, World Health Organization, Geneva.

Yildiz, M., Akdogan, M., Tamer, N. and Oral, B. (2003). Bone Mineral Density of the Spine and Femur . *Calcif. Tissue International*. 72(6): 689–693.

Zipkin, I., Mc Clure, F.J., Leone, N.C. and Lee, W.A. (1958). Relation of the Fluoride Content of Human Bone to Its Chemical Composition. *Public Health Report (Wash)*. 73: 732–740.

6

Impact of Flurosis on Nutrient Intake among School Going Children

Mukta Agrawal, Ritu Babel and Sunil K. Gupta

Endemic fluorosis resulting from high fluoride concentration in groundwater is an important public health problem in India (Kotecha, 2012) as it lies in the geographical fluoride belt that extends from Turkey to China and Japan through Iraq, Iran and Afghanistan (Saravanan et al., 2008). Of the 85 million tonnes of fluoride deposits on the earth's crust, 12 million are found in India (Teotia and Teotia, 1984). A large number of Indians rely on groundwater for drinking purposes and water at many places is rich in fluoride. The main source of fluoride in groundwater is the rocks which are rich in fluoride. Most of the people affected by high fluoride concentration in groundwater live in the tropical countries where the per capita consumption of water is more because of the prevailing climate.

In India, endemic fluorosis is reported from 17 States (fluoride level in drinking water >1.5 mg/l) affecting about 62 million people out of which 6 million are children below the age of 14 years (Susheela, 2001). Toxicity of fluoride adversely affects bones and teeth causing dental and skeletal fluorosis. Apart from this, it also affects many soft tissues of the body especially gastrointestinal system, reproductive system, kidneys, nervous system, etc. known as non-skeletal fluorosis.

Epidemiological observations on the nutritional status of population affected with fluorosis indicate that dietary differences and nutritional imbalances, particularly the low intake of calcium, Vitamin C, Vitamin D and protein, enhances fluoride toxicity and its effect on bone metabolism (Teotia et al., 1981). Fluorosis is further aggravated by malnutrition, the bioavailability of fluoride and the severity of fluorosis is improved in the presence of calcium, vitamin and amino acids deficiencies (Forsyth et al., 1972; Takamori, 1974). The non-skeletal manifestations of fluorosis are often overlooked by people because of wrong notion that fluoride toxicity affects only bones and teeth. Amongst many gastro-intestinal problems, loss of appetite is a common problem. Fluorosis may adversely affect the intake of food and nutrients and, in turn, may be cause of malnutrition among vulnerable sections of the society. The present research work focuses on association of fluorosis with food and nutrient intake by children.

Methodology

The study was carried out on 240 children in the age range of 4–12 years belonging to four villages of Jaipur district, block Sanganer. The selection of the village was done on the basis of fluoride content in drinking water. The fluoride range in drinking water in selected villages were 1.5-<4 ppm, 4–8 ppm, 8–12 ppm and >12 ppm. All the four areas had similar socio-economic and environmental conditions. The rationale for grouping is that usually skeletal fluorosis is not noticed in people consuming water containing up to 4 ppm of fluoride. Severe type of fluoride toxicity is often found in people consuming water containing over 8 ppm of fluoride. In the 4–8 ppm fluoride range a variety of dental, skeletal, and clinical fluorosis has been observed. The 8–12 ppm fluoride range is associated with very severe manifestations. The assessment of dental fluorosis was done by physical examination. Dean's classification (Dean, 1934) was used to categorize severity of dental fluorosis in children.

Dental Fluorosis Grading

Oral Survey for Mottled Teeth: Dean's classification (ibid.) and measurement of effects for clinical diagnosis were used in the study. The degree of severity of mottling was defined by Dean as follows:

Normal Enamel (0): The enamel presents the usual translucent semi-vitriform type of structure. The surface is smooth, glossy and usually of a pale, creamy-white colour.

Questionable Fluorosis (0.5): Slight abrasion from translucency of normal enamel can be seen. It ranges from a few white flecks to occasional white spots. This classification is used in instances where a definite diagnosis of the mildest form of fluorosis is not warranted and a classification of 'normal' not justified.

Very Mild (1): Small opaque, paper white areas scattered irregularly over the tooth but not involving as much as 25 per cent of the tooth surface.

Mild (2): The white opaque areas in the enamel of the teeth are more extensive but do not involve as much as 50 per cent of the tooth.

Moderate (3): The entire enamel surface of the teeth is affected; the surface is subject to attrition and shows wear. Brown stain is frequently a disfiguring feature.

Severe (4): All enamel surfaces are affected and hypoplasia is so marked that the general form of the tooth may be affected. The major diagnostic sign of this classification is discrete or confluent pitting. Brown stains are widespread and teeth often present a corroded-like appearance.

Skeletal fluorosis was assessed using three simple diagnostic tests:

1. The individual was made to bend and touch the toes without bending the knees. If there is pain or stiffness in the backbone, hip and joints, this exercise will not be possible.
2. The individual was made to touch the chest with the chin. If there is pain or stiffness in the neck, this exercise will not be possible.
3. The individual was made to stretch arms sideways, fold the arms and try to touch the back of the head. If there is pain or stiffness in the shoulder joint and backbone, this exercise will not be possible.

Dietary Survey and Estimation of Intake of Nutrients

Interview schedule was used to collect information on dietary data. Food consumption pattern was studied by 24-hour recall method recorded for three consecutive days. It was used to collect information about dietary intake and quantification of nutrients. Help of mothers was taken to recall all foods consumed by the child. To facilitate recall of the amount of food consumed by the child, standard set of utensils (bowls, *bhagonas*, spoons and glasses) were used. Wherever possible cooked samples of *chapaties* were also collected, which were weighed and converted to raw ingredients, according to the standardized data. Nutritive value of food consumed by the children was calculated using food composition tables (Gopalan et al., 2004).

Results

Distribution of Children

Four villages were selected according to their different fluoride levels. The first group consisted of a village with fluoride level in drinking water from 1.5–4 ppm (mg/l) which is named as Group I, whereas Group II comprises of a village having fluoride levels of 4–8 ppm. Group III included a village with fluoride levels (of 8–12 ppm and Group IV had a village where fluoride level was more than 12 ppm).

There were 60 children each in all four groups. More boys (61.66%) have participated in the study as compared to girls (38.33%). There were more boys in group I and II (Table 6.1).

Table 6.1

Distribution of Children on the Basis of Sex

	Male	*Female*	*Total*
Group I	45(75)	15 (25)	60
Group II	43(71.7)	17 (28.3)	60
Group III	34 (56.7)	26 (43.3)	60
Group IV	26 (43.3)	34 (56.7)	60
Total	148 (61.66)	92 (38.33)	240

Note: Percentages in parantheses.

All the children were distributed in three age groups i.e. 4 to 6 years, 7–9 years and 10 –12 years. There were 15 per cent children in the age group 4–6years, 42.91 per cent were in the age group of 7–9 years and 42.08 per cent were from 10–12 years. Maximum (58.3%) children were from 10–12 years of age group in Group III. Least number of children (6.67%) was found in Group II from 4–6 years of age group (Table 6.2).

Table 6.2

Distribution of Children on the Basis of Age

Groups	*Age in years*		
	4–6 yrs.	*7–9 yrs.*	*10–12 yrs.*
Group I (N=60)	17 (28.3)	26 (43.3)	17 (28.3)
Group II (N=60)	4 (6.67)	29 (48.3)	27 (45)
Group III (N=60)	5 (8.33)	20 (33.3)	35 (58.3)
Group IV (N=60)	10 (16.7)	28 (46.7)	22 (36.7)
Total (N=240)	36 (15)	103 (42.91)	101 (42.08)

Note: Percentages in parantheses.

Table 6.3 shows that as the fluoride level in drinking water increases, severity of dental fluorosis also increases. The table depicts that in Group I maximum number of children were suffering from 0.5 and 1 grade of dental fluorosis, whereas in Group IV maximum number of children were suffering from 2 and 4 grade of dental fluorosis.

It is clear from the table that as fluoride content in drinking water increases, prevalence of higher grade of dental fluorosis also increases. In 0.5 and 1 dental fluorosis grades maximum number of children, i.e. 26.66 per cent and 25 per cent of children were from groups that have 1.5–4 ppm of fluoride content in drinking water whereas as fluoride content in drinking water increases up to 8–12 ppm and more than 12 ppm, incidence of higher grades of dental fluorosis, i.e. 3 and 4 grade, increases (Table 6.3).

Few incidences of higher grades can be seen in groups where fluoride content in drinking water is low. Similarly few incidences of lower grades can be seen in higher fluoride content groups. This is due to the fact that as age increases severity also increases because the length of stay in the endemic area also increases. In Group I where fluoride content in water is comparatively low the incidences of higher grades of dental fluorosis was present in higher age groups, i.e. 7–9 and 10–12 years Similarly in Group IV where fluoride content in water is high, incidence of low dental fluorosis grade was found in low age group, i.e. 4–6 years of children.

Table 6.3

Prevalence of Dental Fluososis in Children Belonging to Different Fluoride Levels in Drinking Water

	Dental Fluorosis				
Groups	*Grades of dental fluorosis*				
	0.5	*1*	*2*	*3*	*4*
Group I	16(26.66)	15 (25)	11(18.33)	9 (15)	9 (15)
Group II	10(16.66)	10(16.66)	13(21.66)	7(11.66)	20(33.33)
Group III	7 (11.66)	11(18.33)	8(13.33)	10(16.66)	24(40)
Group IV	2(3.33)	6(10)	14(23.33)	2(3.33)	36(60)
Total	35 (14.58)	42 (17.5)	46 (19.17)	28 (11.67)	89 (37.08)

Note: Percentages in parantheses.

Skeletal Fluorosis

Severe type of fluoride toxicity is often found in people consuming water containing over 8 mg/l of fluoride. Table 6.4 depicts the prevalence of skeletal fluorosis among the selected children. Susheela (2001) also indicated that usually skeletal fluorosis manifestations were noticed in people consuming water containing high range of fluoride. Groups having drinking water fluoride level more than 8 ppm show skeletal fluorosis manifestations.

The signs and symptoms of skeletal fluorosis were found in 30.42 per cent of children. Out of these, 69.6 per cent children did not have skeletal fluorosis symptoms; 23.75 per cent children were in grade 1 while 4.58 per cent and 2.08 per cent children were in grade 2 and 3 respectively. More children from villages having high concentration of fluoride in drinking water, i.e. Group III and Group IV, were found in grade 2 and 3 (Table 6.4).

Along with the fluoride concentration in drinking water, age of children also plays an important role in severity of skeletal fluorosis. Previous studies indicated that at increasing fluoride level of drinking water, skeletal fluorosis occurred at an early age.

Table 6.4

Prevalence of Skeletal Fluorosis in Children belonging to Different Fluoride Levels in Drinking Water

Groups	*Grades*			
	0	*1*	*2*	*3*
Group I	47 (78.33)	13 (21.66)		
Group II	45 (75)	15 (25)		
Group III	48 (80)	11 (18.33)		1 (1.66)
Group IV	27 (45)	18 (30)	11 (18.33)	4 (6.66)
Total	167(69.6)	57 (23.75)	11 (4.583)	5 (2.083)

Note: Percentages in parantheses.

Non-Skeletal Fluorosis

Non-Skeletal manifestations are the early warning signs of fluoride toxicity. It affects the major body systems and is a common type of fluorosis in almost all endemic belts.

Table 6.5

Prevalence of Non-Skeletal Fluorosis in Children belonging to Groups with Different Fluoride Levels in Drinking Water

Clinical Fluorosis	*Grades*		
	0	*1*	*2*
Group I	43 (71.67)	10 (16.67)	4 (6.66)
Group II	44 (73.33)	15 (25)	4 (6.66)
Group III	43 (71.66)	13 (21.66)	4 (6.66)
Group IV	20 (33.33)	11(18.33)	29 (48.33)
Total	150 (62.5)	49 (20.41)	41 (17.08)

Note: Percentages in parantheses.

In the present study, 37.5 per cent children were suffering from clinical fluorosis. Around 20.41 per cent and 17.08 per cent were found in grade 1 and 2 respectively (Table 6.5). More children were found to be suffering from clinical fluorosis manifestations where fluoride concentration in drinking water is high.

Acidity, constipation, pain in the abdomen was seen more frequently in villages having higher fluoride content in drinking water. Recurrent respiratory disorder was also seen in few children exposed to high fluoride in drinking water.

Nutritional Status of Children Residing in Selected Endemic Areas

Epidemiological observations on the nutritional status of population affected with fluorosis indicate that dietary differences and nutritional imbalances, particularly the low intake of calcium, Vitamin C, Vitamin D and protein, enhances fluoride toxicity and its effect on bone metabolism (Teotia et al., 1981). Fluorosis causes loss of appetite, gastro-intestinal problems, poor absorption which adversely affects food and nutrient intake. Poor food and nutrient absorption might cause malnutrition among children.

All the participants were vegetarian. Cereals form the bulk of the diet followed by milk and milk products. Vegetable and pulses were consumed in less quantity vis-a-vis cereals with an almost negligible consumption of fruits. Meals for families were cooked twice a day i.e. in the morning and in the evening. Two-meal patterns were followed by all the children. In the morning they prefer chapatti with buttermilk/caster sugar/jaggery. In the evening any pulses or seasonal vegetable which was available from their fields was preferred. Inadequate diet has also been associated with fluorosis.

The commonly consumed cereals in the study area were pearl millet during winters and wheat in summers. Rice was consumed occasionally. The average cereal consumption was found highest in Group I followed by Group II, III and IV. It was observed that mean intake of cereals and grains decreased in the areas where fluoride level in drinking water was more. Bengal gram and green gram were commonly consumed pulses by the subjects. Average consumption of pulses and legumes among Group I children was 3.59 g where as in Group II children it was 10.18 g which was highest of all Groups. But consumption decreased to 5.34 and 5.23 g in Group III and Group IV respectively (Table 6.6). Consumption of fruits was found almost negligible. Green leafy vegetables and other vegetables were consumed more among children residing in Group II and Group IV as compared to

Group I and Group III. Roots and tubers were also consumed in adequate amount by children. Potatoes and onions were consumed more. Type of vegetables preferred for consumption was dependent on local production and availability of vegetables and fruits with local vendors. Intake of fruits and vegetables was less among Group I children as compared to other groups. Milk was consumed by almost all the subjects in the form of buttermilk, curd or milk as such, or the preparations like kadhi, rabri, etc. Milk and milk products were the main source of calcium. The intake of milk and milk products was less in areas where severity of fluorosis was more. Almost all respondent families had one or another type of milch cattle in their homes and therefore, milk production as well as consumption was high.

Table 6.6

Average Food Intake by Children according to Food Groups in Different Groups

	Group I	*Group II*	*Group III*	*Group IV*
Cereals (gms/day)	208.91	172.96	142.13	118.56
Pulses and legumes (gms/day)	3.59	10.18	5.34	5.23
Fruits and vegetables (gms/day)	53.76	91.96	78.50	85.49
Milk and milk products (ml/day)	441.95	220.64	138.94	100.25
Fats and sugar (gms/day)	31.22	10.82	4.72	4.98

It was found that although they have low cash income but parents prefer to put fat on children's chapattis at least in one meal so that fat consumption was adequate. But in higher fluoride level areas intake of fat and sugars was found to be less.

Nutritional Intake

The intake of energy was mainly through consumption of cereals, pulses, fat and sugar. The average calorie intake of children in all age groups was found to be very low. It was observed that in the areas where fluoride in drinking water was high; mean calorie intake among children was low. The mean energy intake among children from Group I was 1,352.29 kcal which reduced to 929.43 kcal in Group IV children. Mean protein intake also decreased where fluoride level in drinking water was more. In both boys and girls, protein intake also decreased as age increased. The main source of calcium was milk and milk products and its preparations. This probably was the factor which saved the respondents from serious crippling deformities, although the fluoride content of water was high in the study area. Fluoride has an affinity for calcium so it deposits itself on bones and low intake of calcium in the diet relates to the severity of endemic fluorosis while high intake reduces the toxicity of fluoride by forming insoluble salts of fluoride (Singh et al., 1965; Mehra, 1981; Menon, 1985; Gupta et al., 2001). The average calcium intake was found higher in younger age groups (4–9 years) as compared to older ones (10–12 years). Mean calcium intake by children was found decreasing in areas where fluoride level in drinking water was high. In Group I, mean intake was found to be 760.60 mg where as in Group IV it was 391.50 mg (Table 6.7). Iron is a mineral most abundantly found in nature. The main sources of iron in vegetarian diets are cereals and green leafy vegetables. Mean intake of iron was found low in the areas where fluoride content in water was high. Iron intake was found very low (less than 60% of RDA) among children in all age groups.

Highest mean intake of Vitamin A was found among children residing in Group II, whereas lowest was found Group III children. Vitamin A intake was found very low when compared to recommended diatery allowance (RDA) in both the groups. Vitamin C intake was found far below the RDA in all age groups and both sexes. Higher intake was observed in boys as compared to girls in all age groups. The intake of Vitamin C was higher in older children as compared to younger ones. The coexistence of fluorosis and Vitamin C deficiency in the community in an endemic fluorotic area has been reported earlier also. Mean intake of Vitamin C was 19.92 mg in Group I which increased to 34.84 mg in Group II. In Group III it was found to be 23 mg which decreased to 18.92 in Group IV (Table 6.7).

The mean intake of thiamin, riboflavin and niacin also decrease as fluoride content of drinking water increases. In all age groups, thiamine intake was found comparable to RDA. The riboflavin intake was found less than 60 per cent of RDA. The average intake of niacin was also found lower than required amount in all children. Mean fluoride intake from food items was found high. It was seen that niacin intake was decreasing with increase in fluoride levels in water. Krishnamachari and Lakshmaiah (1975) reported that nutritionally inadequate communities suffer more from fluorosis as compared to nutritionally protective groups. Mehra (1981), Susheela (1991) and Gupta et al. (1996) reported that the deficiency of main nutrients such as protein, calcium, and Vitamin C aggravates fluoride toxicity.

Table 6.7

Mean Nutrient Intake by All Children in Different Groups

Nutrients	*Group I*	*Group II*	*Group III*	*Group IV*
Calories (kcal)	1,352.29 ± 289.33	1,257.42 ± 337.96	1,110.89 ± 178.61	929.43 ± 215.37
Protein (g)	40.30 ± 9.07	35.16 ± 9.88	29.86 ± 4.98	23.28 ± 5.64
Fat (g)	39.78 ± 10.77	47.05 ± 15.03	41.78 ± 8.19	36.82 ±12.18
Calcium (mg)	760.60 ± 249.04	586.45 ± 195.69	523.91 ± 131.14	391.50 ± 135.03
Iron (mg)	14.06 ± 3.66	12.04 ± 3.74	9.39 ± 1.87	8.76 ± 2.32
Fat (g)	960.93 ± 516.70	1,203.61 ± 982.68	560.11 ± 213.82	720.43 ± 473.39
Thiamin (mg)	1.16 ± 0.28	1.08 ± 0.34	0.91 ± 0.16	0.75 ± 0.17
Riboflavin (mg)	0.70 ± 0.17	0.62 ± 0.16	0.52 ± 0.12	0.39 ± 0.12
Niacin (mg)	9.83 ± 2.53	9.68 ± 3.14	7.41 ± 1.56	6.23 ± 1.55
Vitamin C (g)	19.92 ± 11.15	34.84 ± 18.76	23.00 ± 12.05	18.92 ± 7.23
Vitamin A (ìg)	12.55 ± 3.09	12.52 ± 3.67	10.93 ± 2.26	8.48 ± 2.06

Nutrient deficiencies have been specifically linked to increased susceptibility to fluoride-induced tooth damage (dental fluorosis), bone damage (osteomalacia), neurotoxicity (reduced intelligence) and mutagenicity. The nutrients of primary importance appear to be calcium, Vitamin C, and iodine, while magnesium, Vitamin D, and protein appear to be important as well.

If we see the other side, the intake of various foods as well as nutrients were found higher in children belonging to villages having low fluoride in drinking water and low food and nutrient intake was observed in children belonging to areas with high fluoride in drinking water. This will further aggravate the malnutrition and micronutrient deficiencies. Fluoride toxicity adversely affects the absorption of nutrients by the intestine` which further reduces the availability of nutrients in body leading to malnutrition. Gupta et al. (2010) also reported that fluoride toxicity causes damage to cells, tissues including gut mucosa which inhibit the absorption of nutrients. Withdrawal of fluoride from ingestion and nutrient supplementation improved BMI and hemoglobin level in pregnant women and improved the birth weight of new born.

Summary

The present piece of research was an attempt to assess the effect of fluoride toxicity on nutrient intake among children. For the study, 240 children of 4–12 years of age were selected from four villages having fluoride content of water in the range of 1.5 to more than 12 ppm. Food and nutrient intake was assessed by 24 hours dietary recall method. Results indicated that severity of dental as well as skeletal fluorosis was higher in children residing in areas where fluoride content of water was high as compared to lower levels. Intake of total energy, protein, calcium, iron, Vitamin A, Vitamin C, etc. were low in children from high fluoride areas as compared to those from low fluoride areas. Further it was observed that the mean nutrient intake as well as per cent of RDA of many nutrients decreased as age of children increased. The possible reason may be reduced appetite due to fluorosis which increased with severity of fluorosis.

References

Arlappa, N., Aalij, Q. and Srinivasa, R. (2013). Fluorosis in India: A Over View, www.ijrdh.com. *Int. J. Res. Dev Health.* 1(2): 97–102.

Forsyth, D.M., Pond, W.C. and Waserman, R.H. (1972). Dietary Calcium Fluoride Interaction in Swine: Effect on Physical and Chemical Bone Characteristic Calcium Binding Protein and History of Adults. *J. Nutr.* 102: 1623–1638.

Gopalan, C., Shastri. D.V.R. and Balasubramanian, S.C. (2004). *Nutritive Value of Indian Foods.* Revised edition. NIN, Hyderabad. 204.

Gupta, S.K., Gupta, R.C., Saxena, A.K. and Gupta, A.B. (2001). Fluoride Toxicity and Fluorosis. Monograph by Applied Research Centre, SMS Medical College. 6.

Gupta, S.K., Gupta, R.C., Seth, A.K. and Gupta, A.B. (1996). Reversal of Fluorosis in Children. *Acta Paediatr Jpn.* 38(5): 513–519.

Gupta, R., Mondal, N.K. and Susheela, A.K. (2010). Guidelines for Practice of International for Rectification of Anaemia and Improving Maternal and Neonatal Heath (MDG 5 and 4). Abstract Volume of XXIX International Society for Fluoride Research, organized by Centre for Advanced Studies Department of Zoology, University of Rajasthan, Jaipur, 2–5 December 2010: 160.

Kotecha, P.V., Patel, S.V., Bhalani, K.D., Shah, D., Shah, V.S. and Mehta, K.G. (2012). Prevalence of Dental Fluorosis and Dental Caries in Association with High Levels of Drinking Water Fluoride Content in a District of Gujarat, India. *Indian J. Med. Res.* 135(6): 873– 877.

Krishnamachari, K.A.V.R. and Lashmaiah, N. (1975). Lack of Effect of Massive Doses of Vitamin C on Fluorids Excretion in Fluoritics During Clinical Trial. *American. J. Clin Nutr.* 28(11): 1234–1236.

Mehra, U.R. (1981). Effect of Ameliorative Measures Against Fluorosis on Certain Blood Constituents of Cattle. *Indian J. Nutr. Diet.* 18: 372.

Menon, R. (1985). Crippling Scourge, *India Today*. 140.

Saravanan, S., Kalyani, C., Vijayarani, M., Jayakodi, P., Felix A., Nagarajan, S., Arunmozhi, P. and Krishnan, V. (2008). Prevalence of Dental Fluorosis Among Primary School Children in Rural Areas of Chidambaram Taluk, Cuddalore District, Tamil Nadu, India. *Indian J Commun Med.* 33: 146–50.

Singh, A., Malhotra, K.C., Singh, B.M., Mathur Khola, O.C. and Khola, H.M. (1965). Endemic Fluorosis: Chemical Constituents of Water. *Indian J. Med. Res.* 53: 866.

Susheela, A.K. (1991). Prevention and Control of Fluorosis: Technical Information for Training-cum-awareness Camp for Doctors, Public Health Engineers, and Others, New Delhi. National Technology Mission of Drinking Water.

Susheela, A.K. (2001). Fluorosis: Indian Scienario: A Treatise on Fluorosis. Fluorosis Research and Rural Development Foundation; New Delhi, India.

Takamori, T. (1974). Recent Studies on Fluorosis. *Fluoride*. 4: 166.

Teotia, S.P. and Teotia, M. (1984). Endemic Fluorosis in India: A Challenging National Health Problem. *J. Assoc. Physicians India*. 32(4): 347–52.

Teotia, S.P., Teotia, M. and Singh, R.K. (1981). Hydro-geochemical Aspects of Endemic Skeletal Fluorosis in India-an Epidemiological Study, *Fluoroide*. 14: 69–74.

7

Prevalence of Dental Fluorosis Due to Fluoride in Drinking Water

A Case Study of Peeplu Tehsil

A.B. Gupta, Kanika Saxena,
Neelam Kothari and Shruti Jain

Fluorine is the thirteenth most abundant element available only in the earth crust. As the groundwater passes through the fluoride rich rocks it carries fluoride with it, hence most of the fluoride is found in groundwater than in surface water. It exists in the form of fluoride in a number of minerals, of which fluorspar; cryolite and fluorapatite are the most common. Fluorosis is an important public health problem in 24 countries, including India, which lies in the geographical fluoride belt that extends from Turkey to China and Japan through Iraq, Iran and Afghanistan. Fluorosis, though a common endemic problem in India, is more widespread and acute in the state of Rajasthan where all the 33 districts have been declared as fluorosis prone. While the WHO standards and IS 10500 (1991) permit only 1.5 mg/l as a safe limit for human consumption, a recent survey of Rajasthan shows that about 24.8 per cent of the 75,266 water samples tested had fluoride level exceeding this limit (Dhindsa, 2006).

Fluorides in drinking water may cause skeletal fluorosis, clinical fluorosis, dental fluorosis or non-skeletal manifestations, while in the final stages it may cause premature aging. Effects on teeth include discolouration, delayed eruption, chipping of edges and pitting etc. as shown in Figure 7.1. So far, the pathophysiology of fluorosis is not well understood. Recent research has resulted in the formulation of

Figure 7.1

Clinical Manifestations of Dental Fluorosis

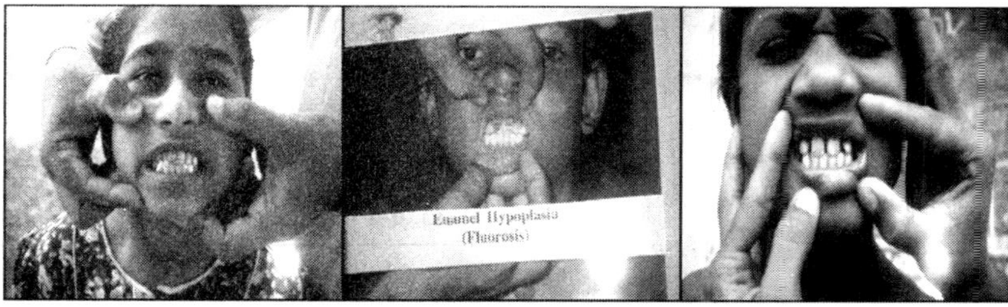

detailed pathophysiology of fluorosis and has also indicated successful treatment of fluorosis (Gupta et al., 1996, 2001, 2012).

The studies attempting correlation between fluoride ingestion and fluorosis manifestation require an objective classification of severity of the disease for establishing any causal relation. Dean (1936) related the appearance and severity of dental fluorosis to different fluoride levels in the drinking water with the aid of a special classification and weighing of the severity of the lesions. This is still the most commonly used index by the researchers on dental fluorosis internationally. The classification is primarily based on the per cent coverage of the stain on teeth. This paper attempts to develop correlation between fluoride ingestion and dental fluorosis manifestation for establishing any causal relation for the data of Peeplu tehsil.

Methodology

Data was obtained from a field study carried out by Public Health Engineering Department (PHED), Rajasthan in the schools of Peeplu tehsil (37 villages) sponsored by UNICEF. For this study the data was collected in the form of number of people observed in each village, number of people affected by dental fluorosis, fluoride levels (0–10.6 mg/l) and total hardness. In this study, 5,583 people were surveyed out of which 764 were affected by dental fluorosis, i.e. 13.68 per cent prevalence of dental fluorosis was observed. Regression analysis was carried out for estimating the relationship between fluoride levels in water and the percentage of people affected by dental fluorosis for all the observations. The observations were further categorized into different ranges and the R^2 was computed for these different ranges for assessing the trend in this value. Regression analysis was carried out for different ranges of fluoride concentration like 0.0–1.0, 0–2, 0–3, 0–5 mg/l, etc. and the percentage of affected people from dental fluorosis. The R square value was noted for all the ranges.

Y (Percentage Affected By Dental Fluorosis)
= A* Fluoride + B (Equation 1)

Where A is regression coefficient and B is the intercept.

For ascertaining the effect of dietary intake from the water (hardness) on the prevalence of fluorosis, multiple regression was carried out in different ranges mentioned above and the R^2 values were further noted.

Y (Percentage Affected By Dental Fluorosis)
= A* Fluoride + B*Hardness + C (Equation 2)

Where A and B are regression coefficients and C is the intercept.

Also, significance testing was carried out at 95 per cent and 50 per cent confidence interval for fraction of population having:

1. 0–1.5 ppm as one data set and above 1.5 ppm fluoride content as the second data set.
2. 0–2.0 ppm as one data set and above 2.0 ppm fluoride content as the second data set.

Results

When the regression test was carried out for all the observations put together it showed a very low R^2 value i.e. 0.044, indicating that the fluoride had no bearing on fluorosis. But a closer analysis of the data indicated that the per cent prevalence were extremely high in some villages. Since no food and nutrition information was known, the effect of intake of Ca and Mg present in water as hardness and the percentage affected by dental fluorosis was assessed by carrying out multiple regression.

Table 7.1 shows the R square values calculated as per the equations 1 and 2.

Table 7.1

Analysis of Peeplu Tehsil

Range	*R2 (Equation 1)*	*R2 (Equation 2)*
0–1	0.311	0.368
0–2	0.024	0.038
0–3	0.048	0.052
0–4	0.021	0.023
0–6	0.053	0.054
0–8	0.019	0.021
0–11	0.044	0.045

R square value obtained was low indicating that only regression analysis is inadequate to derive any relationship in such conditions. It was seen that the R^2 value was obtained highest at 0–1 mg/l indicating the prevalence of fluorosis in 0–1 mg/l fluoride concentration also, not in compliance with standard set by BIS (IS 10500 1991). Figure 7.2 shows the graph of percentage affected by dental fluorosis vs fluoride levels in Peeplu Tehsil.

Figure 7.2

Prevalence of Fluorosis in Peeplu Tehsil

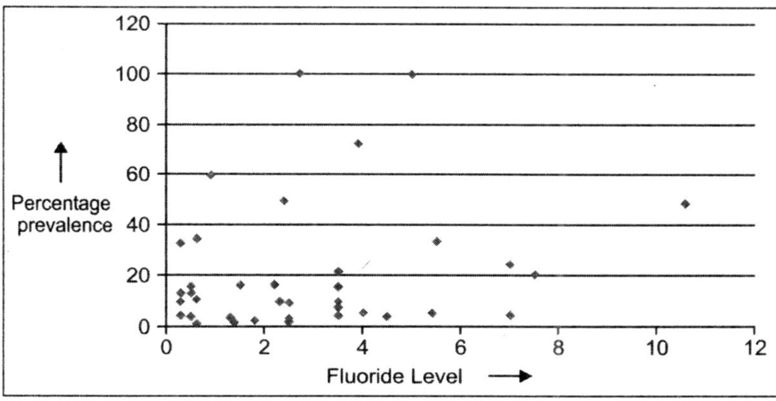

Table 7.2

Significance Testing

Sample Range	*At 50% Interval*	*At 95 % Interval*
PEEPLU (0–1.5 ppm and 1.5 ppm fluoride above)	Significant Difference	Not Significant
PEEPLU (0–2.0 ppm and 2.0 ppm fluoride above)	Not Significant	Not Significant

More comprehensive data about the dietary intakes, age groups, etc. are required. For instance, a case study of Chaksu block, Jaipur showed that the presence of aluminum in drinking water resulted in a significant increase in prevalence of fluorosis and also neurological symptoms such as poor memory and tremors (Gupta et al., 2012).

Conclusion

It can be concluded that fluorosis exists practically in all the villages and is also significant below 1–1.5 mg/l fluoride level in the drinking water which emphasizes the need to revise the existing standard. Also, in epidemiological studies, comprehensive survey is required to derive the cause-effect relationship. Fluoride poisoning can be prevented or minimized by using alternative water sources, by removing excessive fluoride from drinking water and by improving the nutritional status of populations at risk.

References

Gupta, A.B., Sunil K. Gupta, B.S. Jajoo, Kadaverugu, A., Gupta S.K., Gupta, R.C., Seth, A.K. and Gupta. A. (1996). Reversal of Fluorosis in Children, *Acta Pediatrica Japonica*. 38: 513–519.

Dean, H.T. (1936). Chronic Endemic Dental Fluorosis (mottled enamel). *Journal of the American Medical Association*. 107: 1269–1272.

Dhindsa, S.S. (2006). Water Quality Monitoring and Surveillance in Rajasthan, proc. annual convention of IWWA, Jaipur, January 6–8.

Gupta, S.K., Khan, T.I., Gupta, R.C., Gupta, A.B., Gupta, K.C., Jain, P., and Gupta, A. (2001). Compensatory Hyperparathyroidism following High Fluoride Ingestion: A Clinico-biochemical Correlation. *Indian Pediatrics*. 38: 139–146.

Gupta, S.K., Khan, T.I. and Gupta, A.B. (2012). Environmental Health Perspective in Children: Treatment and Prevention of Fluorosis. Lap Lambert Academic Publishing, GmbH and Co., Germany.

IS 10500 (1991). Indian Standard Drinking Water Specification (first revision), Bureau of Indian Standards, India.

World Health Organization (1998). *Guidelines for drinking-water quality*, 2nd ed. Addendum to Vol. 2 Health Criteria and Other Supporting Information. WHO, Geneva, Switzerland.

8

Fluoride Content of Food Items Grown and Consumed in Fluorosis Affected Areas of Rajasthan

Ritu Babel, Mukta Agrawal and Sunil K. Gupta

'Virtually every food contains at least some fluoride. Plants take it up from the water, soil and from the air. In plants fluoride is transmitted through fine hair rootlets into the stems and some reach the leaves' (Waldbott et al., 1978). Intake of high fluoride in diet increases the toxic manifestations of fluorosis. It is also possible that, in areas where fluoride intake via routes other than drinking water (e.g. air, food) is elevated, dental fluorosis develops at concentrations in drinking water below 1.5 mg/l (WHO, 1996). Different plants have different uptakes for example various kinds of tea have an enormous ability to take up fluoride, and that's why many tea products contain 300–400 ppm in dry tea leaves (Xu, 1987). Staple diets rich in sorghum, ragi or bajra containing high silicon besides fluoride seem to aggravate fluoride toxicity in some endemic areas of India (Das, 1996). Cooking food in fluoridated water results in increased dietary fluoride levels. Unlike chlorine, fluoride does not enter the steam when water is boiled. Thus, during cooking the fluoride increases in concentration (ATSDR, 1993). The fluoride content in food material mainly depends upon the following factors:

- Fluoride level in soil and water used for irrigation
- Fluoride level in the atmosphere and

- Use of fertilizers, insecticides and other sources of contamination.

The general impression so far has been that endemic fluorosis is due to ingestion of water with high fluoride levels. There is now disturbing evidence that food-borne fluorosis is emerging as a major possibility. In China, for example, in the county of Guizhon, staple food items, namely, rice, corn, cabbage, soya-bean, potatoes and wheat have been found to have fluoride content ranging from 8.3 to 11.7 mg/kg; it is now recognized in China that both water and food could contribute to fluorosis. So the routine estimation of fluoride levels in foods must now be included as part of the systematic investigation (Susheela, 1984).

When considering fluoride recommendations on a group basis, all major sources of fluoride ingestion, such as diet, dentifrices, and fluoride supplements, must be considered so that total daily intake of fluoride can be estimated. To this end, efforts should be renewed to determine and regularly monitor the fluoride levels of beverages and foods and have their fluoride levels made available to the profession and the public (Levy et al., 1999). Estimation of the amount of fluoride ingested from all environmental and dietary sources is important so that balanced and scientifically sound decisions can be made.

Although it is very well recognized that fluoride enters body through water and various food items; the data available in the country, regarding food items, is rather scanty. The information that is available is based on food analysis or agriculture crop analysis of certain regions/locations. The data thus emerging from one particular location is unlikely to be valid for other regions or locations. The main reason is the fluoride content of water and soil being variable from location to location. There is a need for fluoride analysis of agricultural crops district-wise, which will add information to educate and create awareness on the importance of selection of agricultural crops which are health-promoting foods and crops.

Materials and methods

Sample Collection: Food samples were collected from the nearby villages of Jaipur district where fluoride in water ranges from 2–12 ppm. Villages were selected on the basis of fluoride content in drinking water and incidences of dental fluorosis. Possible natural sources of fluoride for human living in these endemic areas are drinking water and other food items. Therefore, samples of vegetables and grains cultivated in agricultural lands of the villages were collected for analysis.

Sample Material: Due to a drought situation in Rajasthan very few crops were cultivated in the area as sufficient water for irrigation was not there. Around 48 samples were collected from endemic areas for analysis of fluoride content. Mainly bajra and wheat are cultivated in the areas as they constitute the main diet. Some seasonal vegetables were also grown for own use. All the vegetables and grains grown or available in the endemic areas for consumption were collected and were analysed for their fluoride content. Very few vegetables which were not grown in that particular area came from nearby villages which also fall under the endemic zone. Besides vegetables and grains, milk samples were also analysed for their fluoride content. Generally cow and goat milk was consumed mainly by younger children in the study areas. Buffalo milk was consumed only by elder ones.

Locally available tea leaves were also analysed. Generally two types of tea leaves were available and consumed by the villagers – open and packed one. In the packed only one brand i.e. Maheshwari pack, was generally consumed in the villages under study.

Method for Estimation: Potentiometric method was used for estimation of fluoride in food items (AOAC, 1995). In this, fluoride is extracted from dry samples with HNO_3 followed by aqueous KOH. Slurry is adjusted to pH 5.5 and complexing agent and background fluoride is added. Calibration curve was prepared using standard solutions 0.1, 0.2, 0.5, 2 and 10 ppm. Readings were recorded in mV for these concentrations and graph was prepared on semi-log graph paper. For every mV concentration C was calculated by curve. Fluoride was calculated by using:

$$F\ (\mu g/g) = (C - 0.10) \times 50/w$$

Where C = ppm F from curve, 0.10 = ppm background F in final solution, 50 =ml final solution, w = g sample.

Sensitivity and Reproducibility of Method used for Estimation: To check the method, a known amount of sample was tested. So that it can be judged that how much sensitive the results are from the particular method. One sample was tested 3–4 times to check the repeatability of results. All the samples were analysed thrice. Several other methods were also examined. Based on the laboratory studies, Potentiometric method (AOAC, 1995) gave acceptable results and was adopted for analysis of common food items consumed in rural areas.

Results and discussion

Well water is used generally for irrigation purpose. The underground water was mostly free of bacteria but it should be tested for its natural chemicals, such as fluoride. Flouride is widely present in rocks from where the water is pumped. Failure of rains has caused groundwater levels to drop and fluoride levels to concentrate alarmingly. Fluoride in water samples in this endemic area ranged from 2–12 ppm.

The average values of fluoride estimated in foods grown or locally available and consumed are given in Table 8.1.

Table 8.1

Mean Fluoride Content in Raw Foods

Food Items	*Fluoride content, dry weight (ppm)*	*Fluoride content, wet weight (ppm)*
Cereals		
Maize	36 ± 0.20	30.6
Wheat flour	25.6 ± 2.11	22.47
Bajra	25.4 ± 1.13	22.25
Pulses and legumes		
Bengal gram	25.6 ± 2.01	23.04
Green gram dal	36 ± 0.56	32.25
Masoor dal	25.6 ± 1.18	22.42
Red gram dal	11.6 ± 1.11	10.04
Peas	36 ± 0.98	9.75
Cowpea	23.6 ± 1.08	20.43
Leafy vegetables		
Cabbage	79 ± 0.56	6.39
Radish leaves	70.4 ± 0.30	6.47
Spinach	37.2 ± 0.36	2.93
Coriander leaves	44.2 ± 1.35	6.05
Bathua leaves	21.2 ± 0.36	2.204
Cowpea leaves	30.6 ± 0.66	3.36
Fenugreek leaves	11.1 ± 0.53	1.54
Other vegetables		
Cucumber	22.4 ± 0.72	0.82
Cluster beans (Guar)	22.6 ± 1.18	4.29
Tomato	32.2 ± 0.56	1.93
Brinjal	66.4 ± 0.62	4.84
Ladies' Finger	55 ± 0.72	5.5
Cauliflower	40 ± 0.61	3.6
Ridge gourd (Torai)	6.8 ± 0.62	0.326
Kachari	47.2 ± 0.70	-
Bottle gourd (Lauki)	31.2 ± 0.87	1.21

Cont'd...

...Cont'd

Field beans (Sem)	22.6 ± 3.8	3.14
Tinda	21.6 ± 2.62	1.4
Bitter gourd (Karela)	15.6 ± 1.15	1.18
Tomato (Green)	44.2 ± 3.76	3.049
Roots and tuber		
Radish	57 ± 1.67	3.19
Carrot	42.8 ± 2.25	5.99
Potato	27.8 ± 1.25	10.82
Onion	23.6 ± 1.08	3.16
Beverages		
Tea leaves (Maheswari pack)	144.8 ± 1.40	-
Tea leaves (open)	123.4 ± 1.57	-
Spices and condiments		
Green chilly	34.8 ± 2.26	4.97
Red Chilly powder	42.8 ± 0.87	38.52
Coriander powder	41.4 ± 1.35	36.52
Cumin seeds	26.6 ± 1.61	23.43
Garlic	18 ± 1.71	6.84
Ginger	36 ± 0.62	6.87
Turmeric	58.8 ± 0.85	51.09
Ajwain	70.4 ± 1.93	65.19
Fenugreek Seeds	16 ± 1.15	13.8
Milk and milk products		
Milk (Cow)	-	13 ± 0.72
Milk (Buffalo)	-	12.2 ± 0.62
Milk (Goat)	-	9.8 ± 0.79
Butter Milk	-	8.5 ± 0.53

The fluoride content of cereals ranged from 25 to 36 ppm. Among the grain samples analysed, the fluoride content in bajra and wheat flour was 25.4 and 25.6 ppm, whereas in maize it was found to be 36 ppm. For pulses and legumes used in the endemic areas, fluoride content ranged between 11.6 ppm to 36 ppm. Among the pulses, red gram dal was found to have lowest fluoride content (11.6 ppm) and green gram dal had highest fluoride content (36 ppm). The other samples Bengal gram dal, lentil, cowpea had fluoride content of 25.6 ppm, 25.6 ppm and 23.6 ppm respectively.

Out of 7 leafy vegetables analysed, cabbage had highest fluoride (79 ppm) content whereas lowest (11.1 ppm) was observed in fenugreek leaves. Radish leaves also contained high amount of fluoride i.e. 70 ppm. Other samples: coriander leaves, spinach, cowpea leaves and bathua leaves had fluoride content 44.2 ppm, 37.2 ppm, 30.6 ppm and 21.2 ppm, respectively.

Among the other vegetables analysed, the fluoride content ranged from 6.8 to 66.4 ppm. Brinjal had highest fluoride content (66.4 ppm) and ridge gourd had lowest (6.8 ppm) fluoride content. The other samples of cucumber, cluster beans, fieldbeans and tinda had fluoride contents of 22.4 ppm, 22.6 ppm, 22.6 ppm and 21.6 ppm, respectively. Tomato, kachari, bottle gourd and green tomato contains 32.2 ppm, 47.2 ppm, 31.2 ppm and 44.2 ppm, respectively. Cauliflower and ladies' finger had 40 ppm and 55 ppm fluoride contents, respectively. Bitter gourd had 15.6 ppm fluoride. Out of 4 roots and tubers samples analysed, radish had highest amount of fluoride (57 ppm). Carrot, potato and onion had 42.8 ppm, 27.8 ppm and 23.6 ppm of fluoride contents, respectively.

Among spices and condiments, fluoride content ranged between 16 to 70.4 ppm. Fenugreek seeds had lowest (16 ppm) ajwain had highest (70.4 ppm) and cumin seeds had 26.6 ppm of fluoride content. Green chillies, ginger and garlic had fluoride content of 34.8 ppm, 36 ppm and 18 ppm, respectively whereas red chilly powder, turmeric powder and coriander powder had 42.8, 58.8 and 41.4 ppm of fluoride contents, respectively.

Two samples of dried tea leaves, commonly used in endemic areas were also analysed for their fluoride contents. The one packed tealeaves (Maheshwari chai) contained 144.8 ppm and the other tea leaves sold as open tea leaves contained 123.4 ppm of fluoride.

Among the milk and milk products samples analysed fluoride content ranged between 8.5 to 13 ppm. Of the three milk samples analyzed the cow's milk had 13 ppm of fluoride, whereas in buffalo's and goat's milk it was 12.2 and 9.8 ppm. The sample of buttermilk contained 8.5 ppm of fluoride.

Fluoride content in different food items was studied earlier by different investigators. The available data indicates that data from a location cannot be compared with data from another region or location, because the fluoride content of water and soil varies from location to location. For example Chari et al. (1974) reported only 2.8 ppm fluoride content in bajra where fluoride in water ranged from 2.5–5 ppm, at Podli, Andhra Pradesh. Whereas Rajyalakshmi (1982) observed very high content of fluoride (74 ppm) in bajra and that in water ranged from 1–12 ppm (Andhra Pradesh). Similarly, Lakhadwala and Punekar (1973) reported fluoride in pulses and water ranging between 3.84–4.84 ppm and 0.15–0.48 ppm, respectively (Bombay), whereas Rajyalakshmi (1982) reported fluoride in pulses ranging between 14.8 to 52.8 ppm.

The fluoride content of milk is higher than that of water. The difference between cow, buffalo and goat milk may be due to difference in the quantity of water and vegetation consumed. The physiological mechanism which determines the composition of milk also plays a role. In its pure form milk is a significant source of fluoride to man. Since the more susceptible age for dental fluorosis is up to 14 years and the consumption of milk during this age is high, fluoride from milk may account for the increase in the incidence of fluorosis. But at the same time milk is a rich source of calcium too. Calcium plays an important role in diminishing the effect of fluorosis. So the high fluoride in milk doesn't actually affect children.

Variations observed in fluoride content of vegetables and grains may be attributed to variation in fluoride content of irrigation water and the accumulation efficiency of the plant.

Earlier studies in these areas depict that people are consuming high levels of fluoride (2–12 ppm) in drinking water and are suffering from dental, clinical and skeletal fluorosis. Studies on distribution of high fluoride groundwater and prevalence of fluorosis have indicated that substantial number of cases of dental fluorosis and non-skeletal manifestations were identified. And in higher fluoride levels in drinking water, cases of skeletal fluorosis were also seen.

According to the dietary pattern of people residing in these areas it was found that all the families were vegetarian. Cereals form the bulk of the diet followed by milk and milk products. Vegetables and pulses were consumed in lesser quantity than cereals with almost negligible consumption of fruits.

In cereals, bajra and wheat are mainly used. Chapaties made from bajra flour constituted the staple diet. Wheat flour is used generally in summers. In addition to chapaties, rabri and bajra khich were also mainly consumed. Vegetables and pulses were consumed generally for dinner. Any seasonal vegetable which is available from their fields is preferred. These families belong to a low income group i.e. limited cash income, vegetables and pulses are therefore seldom bought from the market.

Thus cereals and vegetables grown in these areas, with high fluoride groundwater, also contain some amount of fluoride. It will further aggravate the prevalence of fluorosis.

Though the fluoride content of vegetables and grains vary, however it is evident that in whatever combination they are combined, a significant amount will enter the human body.

Conclusion

The role of drinking water in the incidence of fluorosis is well recognized. Therefore, it is essential that the villages affected by fluorosis are supplied with safe drinking water. Since vegetables, grains and milk and other food products are also significant sources of fluoride, their fluoride content should also be taken into account.

In endemic areas, along with analysis of drinking water, the analysis of fluoride content of agricultural crops should also be carried out which will add information to educate and create awareness on the importance of selection of agricultural crops which are low in fluoride content. In areas where water with high fluoride content is used, the intake can be several times greater. We must take note of the fact that irrigation programmes and new intensive agricultural technology are bringing about striking changes in soil chemistry and subsoil water composition; the impact of these developments on the fluoride levels of foods in endemic areas needs to be monitored rigorously. More investigations should be carried out to ascertain various source of fluoride for people living in the endemic areas.

References

AOAC (1995). Official Methods of Analysis, Association of Official Analytical Chemists, 14th edn, Washington, DC. 58: 1129

ATSDR (1993). Toxicological Profile for Fluorides, Hydrogen Fluoride and Fluorine, Agency for Toxic Substances and Disease Registry, US Department of Health and Human Services, Public Health Service.

Chari, K.V, Rao, R.J. and Naidu, M.G.C. (1974). Fluorine Content of Raw Vegetables, Foods Available at Podili, Andhra Pradesh. *Proc Symp on Fluorosis*, October 1974, Indian Academy of Geoscience, Hyderabad, India. 144

Das, A. (1996). Fluorosis. *Textbook of Nutrition*. Bamji M.S., Rao N.P., Reddy V. (eds) Oxford & IBH, New Delhi. 434–435.

Lakhadwala, D.R. and Punekar B. D. (1973). Fluoride Content of Water and Commonly Consumed Foods in Bombay and A Study of Dietary Intake. *Indian J Med Res.* 16: 1679–1687

Levy, S.M. and Guha Chowdhury, N. (1999). Total Fluoride Intake and Implications for Dietary Fluoride Supplementation. *J Pub Heal Dent.* 59: 211–23.

Rajyalakshmi, K. (1982). Technical Report, Department of Environment, Ministry of Environment and Forests, Government of India.

Susheela, A.K. (1984). Epidemiology and Control of Fluorosis in India. Bulletin of Nutrition Foundation of India.

Waldbott, G.L, Burgstahler, A.W. and McKinney, H.L. (1978). *Fluoridation: The Great Dilemma*, Coronado Press, Inc. Lawrence, Kansas.

WHO (1996). *Health Criteria and Other Supporting Information: Guidelines for Drinking-water Quality*, 2nd Edn, World Health Organization, Geneva. 2: 231–237.

Xu, C.C. (1987). The Study on the Ability of Selectively Accumulating Fluorine in Tea Leaves. *Food Sci.* 5: 10–12.

Section–III
Nitrate, Heavy Metals and Salinity

9

Environmental Health Aspects of Nitrate Toxicity in Humans and its Mitigation Measures

Sunil K. Gupta

Naturally occurring nitrate levels in surface water and groundwater are generally a few milligrams per litre. Many a times, in groundwater higher levels of nitrate are found due to water percolating through nitrate rich rocks and also due to excessive use of chemical fertilizers. World Health Organisation (WHO, 2004) maintains that extensive epidemiological data support limiting the value of nitrate – nitrogen to 10 mg/l or as nitrate to 50 mg/l for human consumption, whereas IS–10500 prescribes maximum permissible limits of nitrate in drinking water as 45 mg of NO_3 per litre (IS 10500, 1995).

The problem of nitrates is endemic internationally, as well as nationally but so far no compiled data at national level are available. The data available are either of small areas or of scattered zones/areas. Stray studies tried to compile status of nitrate in drinking water in Rajasthan State, India (Kumar et al., 2002; PHED Habitat Survey, 1991–93).

Epidemiology

The problem of nitrate toxicity is global and has been reported from a number of countries (Prakasa Rao and Puttanna, 2006).

International Perspective

Nitrate is a widespread contaminant of ground and surface water worldwide (Prakasa Rao and Puttanna, 2006; Singh et al., 1994). In the United States, the problem appears to be concentrated in the Mid-West and the Far-West, with large areas of Iowa, Illinois, Kansas, Michigan, Wisconsin, Washington and California being heavily affected (Revenga and Mock, 2001; Smith et al., 1994). In South America, nitrate concentration in the monitored waterheds is relatively low and follows human land use. The highest nitrate concentration is found in the Uruguay waterhed, where intensive agriculture is practised. Likewise, nitrate concentration in water is also greater in the Magdalena waterhed of Colombia than in the less densely populated waterheds of the Amazon basin. The nitrate concentration in South America corresponds to lower fertilizer application rates in comparison to Europe.

Nitrate concentration are higher in waterheds that have been intensively used and modified by human activity, such as the Weser, Seine, Rhine, Elbe and Senegal. High levels are also found in such waterheds in China, South Africa and the Nile and Mississippi basins. South Africa has some of the highest natural nitrate levels in the world (more than 550 mg/l NO_3-N) especially in the Kalahari region (Colvin et al., 2008). More detailed data are available in Europe and show distinct regional trends in the concentration of nitrates in rivers (Scheidleder et al., 1999). Nitrate loadings are highest in areas with intensive livestock and crop production, especially in the northern parts of Western Europe. Nitrate concentration is lowest in Finland, Norway and Sweden.

National Perspective

Nitrate from sources like mineralized organic matter, fertilizers, organic manures, industrial effluents or urban wastes, results in groundwater pollution which has received attention in various parts of India. The available data from various studies point to the fact that nitrate pollution from any of the above sources can pose a serious health hazard to humans, animals, birds and aquatic life. The data available on concentration of nitrate in groundwater samples from Punjab, Haryana, Uttar Pradesh, Delhi, Orissa (now Odisha), Tamil Nadu, Bihar, Andhra Pradesh, Madhya Pradesh, Maharashtra, Karnataka and other states point to the fact that in many of these samples, the nitrate content in groundwater has been much more than permissible limits. In many parts of Punjab and Haryana nitrate levels in groundwater is continuously increasing beyond the critical standard. Nitrate levels in groundwater over vast agricultural areas can be correlated with intensive irrigated agriculture, use of nitrogenous fertilizers and groundwater exploitation. Studies tried to evaluate the nitrate pollution and the strategies to reduce it (Prakasa Rao and Puttanna, 2006).

Sources

There are two sources of nitrogen nitrate for human beings:

1. Exogenous sources.
2. Endogenous sources.

Exogenous Sources

Wastes containing organic nitrogen are decomposed in soil or water by microbial action forming ammonia, which is then oxidized to nitrite and nitrate. Because nitrite is easily oxidized to nitrate, it is primarily nitrate that is found in groundwater and surface water. Nitrate-containing compounds in the soil are generally soluble and readily migrate with groundwater. Nitrosamines have a short lifespan in ambient air. Vegetables (cauliflower, spinach, collard greens, broccoli and root vegetables) account for more than 70 per cent of the nitrates ingested in human diet. The remainder of nitrate in a typical diet comes from drinking water (21%) and meat and meat products (6%). Nitrates are naturally occurring inorganic ions, which are part of the nitrogen cycle and are commonly found in groundwater and surface water. The nitrate ion is the most stable form of nitrogen in oxygenated environments, thus all nitrogen-containing molecules can act as sources of nitrates.

Water and Food

Main sources contributing to nitrate content of natural water are atmosphere, geological features, anthropogenic sources, atmospheric nitrogen fixation and soil nitrogen. Oxides of nitrogen are generated through lightning and reach to ground with rain water. Recent reports indicated that atmospheric contributions amount to 25 per cent of total load of nitrate (WHO, 2003). Direct discharge to septic tanks, sewage and industrial effluent are other contributors. Excessive use of chemical fertilizers is one of the main sources of nitrate in water.

Nitrogen is an essential major element for plant growth. Nitrogen nutrition of crops is very often the limiting factor for food production and nutritional quality of food as indicated by its protein content. Thus, nitrogen has a key role to play in feeding the world population. Since a large response to nitrogenous fertilizers is a universal phenomenon, farmers usually apply large doses of fertilizer. But the recovery of applied nitrogen by the crop rarely exceeds 50 per cent and a large fraction of the surplus nitrogen gets transformed into nitrate form and pollutes the environment including surface and ground water (Figure 9.1).

Although there has been a concurrent rise in fertilizer usage with nitrate concentration in water, particularly where irrigation is practiced, there exist several other sources of nitrate-N, namely sewage effluent, animal excreta, natural soil nitrate and decomposition of soil organic matter. In many regions, natural deposits of nitrate may constitute a major source for nitrate pollution of groundwater. For example, in selected locations in Punjab, Haryana and western Uttar Pradesh, nitrate-N concentration in well water samples was several folds higher than the upper safe limit.

Figure 9.1

Nitrogen in Agricultural System

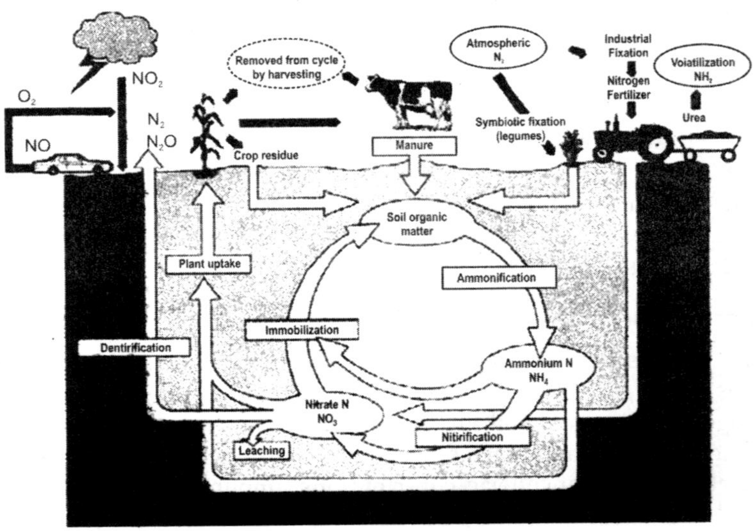

Leafy Vegetables

Being a rich source of nutrients and antioxidants, leafy vegetables occupy an important place in human diet. Some vegetables are known to accumulate high concentration of nitrate under the current practice of heavy fertilization. Vegetables are the major source of the daily intake of nitrate by human beings, supplying about 72 to 94 per cent of the total intake. On consumption by humans, part of this nitrate is converted to nitrite and N-nitroso compounds that have detrimental effects on human health.

In 1995, the European Commission's (EC's) Scientific Committee for Food (SCF) established the Acceptable Daily Intake (ADI) of nitrate ion as 3.65 mg kg^{-1} body weight (equivalent to 219 mg day^{-1} for a person weighing 60 kg) (Scientific Committee on Food, 1995) and the Joint Expert Committee on Food and Agriculture (JECFA) in the United Nations/WHO established the acceptable daily intake (ADI) of nitrate as 03.7 mg kg^{-1} body weight (Speijers, 1996). Therefore, assuming a 60 kg body weight, the ingestion of 100 g of fresh vegetables with a nitrate concentration of 2500 mg kg^{-1} fresh weight exceeds the Acceptable Daily Intake for nitrate by approximately 13 per cent. For a real assessment, however, nitrate content in all other sources as well as their average daily consumption amount needs to be considered. On the other hand, the USA Environmental Protection Agency (EPA) Reference Dose (RfD) for nitrate is 1.6 mg nitrate N kg^{-1} body weight per day (equivalent to about 7.0 mg NO_3 kg^{-1} body weight per day) (Mensinga et al., 2003).

The nitrate content in vegetables available in the markets is reaching as high as 4451 mg kg^{-1} fresh weight in chenopodium and 4293 mg kg^{-1} fresh weight in spinach (Anjana et al., 2007a; Anjana and Iqbal, 2007b). Studies on genotypes of spinach have shown enormous intraspecific variation of nitrate content. One genotype at the three-week stage of plant growth and six genotypes at the six-week stage exceeded the ADI limit. Petioles possessed several times higher level of nitrate than leaf lamina. All of the genotypes studied showed diurnal variation in nitrate accumulation with a minimum at noon (ibid.).

Studies suggest that nitrate content possesses a significant reverse relationship with nitrate reductase (NR) activity, the first enzyme in nitrate assimilatory pathway. It is held that NR expression level is largely responsible for different nitrate accumulation patterns in different cultivars of a given species. Therefore, nitrate accumulation in plants can be significantly reduced by over expression of NR genes in high nitrate-accumulating genotypes. In addition, it is also important to enhance expression of NR so that nitrate does not accumulate in the green tissue after converting into nitrite. Overexpression of these genes will reduce the nitrate concentration in vegetables to safe limits for human consumption. Moreover, a careful selection of vegetable genotypes based on their relative levels of nitrate content and NR activity, harvesting young plants at noontime and removal of petioles from leaves could minimize the dietary intake of nitrate through leafy vegetables. Also, at least 50 per cent of accumulated nitrate can be removed by cooking vegetables in water (with low nitrate concentration).

Other Sources of Nitrate Pollution

Malik (2000) suggested that the presence of nitrate bearing rocks in geological formations as in some parts of Haryana contributed to high nitrate levels in groundwater (Table 9.1).

Table 9.1

Other Sources of Nitrate Pollution

	Agriculture	*Municipal*	*Industrial*
Diffused sources	Use of synthetic nitrogen fertilizers	Combustion engines in vehicles	Atmospheric emission (oxides of nitrogen) from combustion engines and energy production
	Use of organic manures	Disposal of municipal effluents	Disposal of effluents
Point sources	Accidental spills	Badly designed land-fills	
	Improper shelter less composting pits, heaps	Septic tanks	Disposal of wastes into wells, ponds, injection into tube wells, etc.
		Leaking sewerage Systems	Badly designed landfills
Linear sources	Contamination of rivers with major groundwater connections		

Food including meat and meat products and vegetables account for 76 per cent and drinking water accounts for 21 per cent of total human nitrate intake (WHO, 1985b). Vegetables and cured meat are the main sources of nitrate and nitrite in diet, but small amounts can be present in fish and dairy products. Nitrate is added to meat as a preservative and to ensure better colour and flavour. In general, vegetables will be the main source of nitrate intake when nitrate levels in drinking water are below 10 mg per litre (WHO, 1985a). Several vegetables and fruits contain 200 to 2500 mg of nitrate per kg (Chaudhary et al., 2010). The nitrate content in vegetables can be affected by processing of the food, the use of fertilizers and growing conditions, especially the soil temperature and daylight intensity. Vegetables such as lettuce, spinach, beetroot, celery, brinjal (eggplant), beets, banana, strawberry, tomatoes and peas (Paliwal, 1989) often contain nitrate concentration above 2,500 mg/kg, especially when they are cultivated in greenhouses. Nitrate levels in food are generally well below 10 mg/kg and rarely exceed 100 mg/kg. Exceptions to this are vegetables that have been damaged, poorly stored, or stored for extended periods as well as pickled or fermented vegetables. In such circumstances, nitrate levels of up to 400 mg/kg have been found (Sinha et al., 2008).

Preservatives used in food industry, e.g. for preservation of meat and fish, nitrite for meat curing, are significant nitrate sources. Cooking in aluminum utensils enhances reduction of nitrates to nitrite (WHO, 1977e) hence increasing the risk.

Continuous application of nitrogen fertilizers can cause an increase in the concentration of nitrates in fruits, vegetables and food. Maximum admissible level of nitrites in vegetables is 1 mg/kg (Mondal et al., 2008). Excessive use of animal manures can lead to high nitrate content in vegetables. It is necessary to analyse nitrate content in organically grown fruits, vegetables and other food articles.

Hydro-geological investigations show that nitrates levels are high in sandy soil because of low water holding capacity and high permeability of pollutants like chloride and nitrate than in clayey soil.

Top feed prepared with water containing high nitrate also exposes the infants to high nitrate. As nitrate levels in breast milk are low, the probability of exposure to high nitrate in breastfed infants is very low.

Endogenous sources

Nitrates known as endogenous nitrate are also produced in the body. A major pathway for endogenous nitrate production is conversion of arginine by macrophages to nitric oxide and citrulline, followed by oxidation of the nitric oxide to nitrous anhydride and then reaction of nitrous anhydride with water to yield nitrite. Gastrointestinal infections and non-specific diarrhea increase endogenous (non-bacterial) nitrate synthesis, probably induced by activation of the mammalian reticuloendothelial system (WHO, 1985b; WHO, 1996).

Total Daily Intake of Nitrate (WHO, 1985c)

In general, diet is the major source of human intake of both nitrate and nitrite. Certain vegetables, e.g. lettuce, spinach, beetroot and celery, contain relatively high levels of nitrate (often over 226 mg/kg), but the nitrite levels are usually very low. Nitrates and nitrites are also added as preservatives in some foods, such as cured meats and certain types of cheese.

In most European countries, the mean nitrate intake is about 10–30 mg/day. Vegetarians usually have a two to four fold higher intake of nitrates than non-vegetarians.

For most people in Europe, drinking water does not contribute more than 30 per cent of their total dietary intake of nitrate. However, when the nitrate level in drinking water exceeds 10 mg/l, this contribution may be considerably higher.

For bottle-fed infants, drinking-water is usually the major source of dietary nitrate. Nitrate levels in breast milk are low, even when the lactating mother consumes nitrate-rich drinking water. Breast-feeding should thus be encouraged, even in areas with high nitrate levels in drinking-water.

Predisposing Factors

A direct correlation between gastric pH, bacterial colonization and gastric nitrite concentration has been observed in healthy people (Mueller et al., 1986). In individuals with gastrointestinal disorders and achlorhydria, high levels of nitrite have been reported (Ruddell et al., 1978; Dolby et al., 1984). Infections and non-specific diarrhea played a role in the increased endogenous synthesis of nitrate (Green et al., 1981; Hegesh and Shiloah, 1982; Bartholomew and Hill, 1984; and Gangolli et al., 1994). These observations are all consistent with the induction of nitric oxide synthase by inflammatory agents. This induction in humans has been difficult to demonstrate directly, but administration of (15_N) arginine to two volunteers resulted in the incorporation of 15_N into urinary nitrate in both individuals, confirming the arginine-nitric oxide pathway in humans (Leaf et al., 1989).

Chronic Toxic Effects

Human lethal doses of 4–50 g NO_3 (equivalent to 67–833 mg NO_3/kg BW) have been reported. Toxic doses with methemoglobin formation as a criterion for toxicity – ranged from 2 to 5 g (Corré and Breimer, 1979) of NO_3. These values are equivalent to 33–83 and 100–150 mg NO_3/kg body weight respectively. Long-term exposure of non-lethal doses causes chronic toxic effects. The following effects have been reported:

Methaemoglobinemia

Nitrates in drinking water have been reported to cause Methaemoglobinemia in infants up to 6 months of age.

Mechanism of Methemoglobin Formation

The essential action in the formation of methemoglobin is an oxidation of the ferrous to the ferric ion. This oxidation may be brought about in one of the following ways (Bodansky, 1951): by the direct action of the oxidants, or by the action of hydrogen donors in the presence of oxygen, or by autoxidation. In the presence of nitrites, the ferrous ion of haemoglobin gets directly oxidized to ferric state. Normally, the methemoglobin that is formed is reduced to haemoglobin by cytochrome b_5/cytochrome b_5 reductase system, as the following reaction (Benz, 2001):

Hb^{+3} + Red cyt b_5 $\rightarrow$ Hb+2 + oxy cyt b_5

Oxidised cytochrome b_5 (Red Cyt b_5) is regenerated by the enzyme cytochrome b_5 reductase:

cyt b_5

oxy cyt b_5 + NADH $\rightarrow$ Red cyt b_5 + NAD Reductase

Cytochrome b_5 reductase is a NADH dependent enzyme. NADH is supplied from glycolysis (Figure 9.2).

Figure 9.2

Reduction of Methemoglobin (Nagel, 2007)

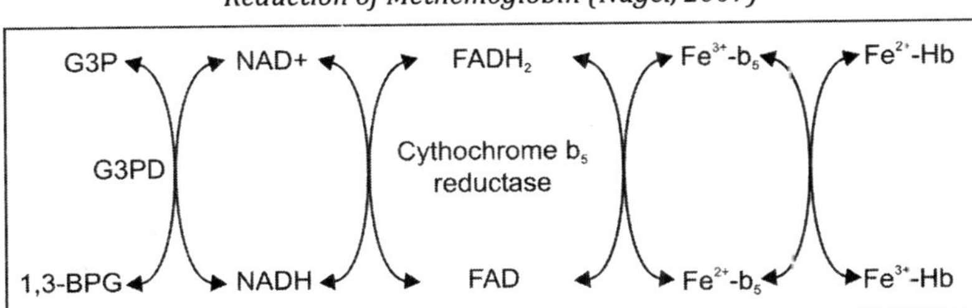

Thus, the enzyme cytochrome b_5 reductase plays a vital role in counteracting the oxidative effects of nitrate ingestion. However, permissible concentration (50 mg/l as nitrate as per WHO, 2004) normally present in water or food do not cause any health risk to adults but the infants constitute a vulnerable group for the following reasons (WHO, 1977b):

- Relatively higher stomach pH (2.0–5.0) which permits growth of nitrate reducing organisms such as *Coliforms, E. coli, Pseudomonas fluorescence, B. subtilis, Staph. Albus*

- Relative higher consumption of water per unit weight of body
- The presence of foetal Hb which readily gets oxidized to methHb
- Poorly developed cytochrome b_5 reductase system
- Nitrate gets concentrated by repeated boiling of water for feeding
- Bacterial contamination of the water itself or dried milk powder
- Early weaning on to nitrate rich vegetables, e.g. spinach
- Diarrheo causes increase in stomach pH

In India where the breast-feeding is a common practice, especially up to the age of 6 months, protects the infants from nitrate toxicity. The probability of toxicity to foetus in pregnant women cannot be ruled out as transplacental passages of nitrate metabolite (Nitrite) have been documented in animal experiment. While a few cases of methaemoglobinemia in infants have been reported to be associated with water nitrate levels of less than 50 mg nitrate ion/L, most cases occur with nitrate level of 90 mg nitrate ion/L or more (Comly, 1945; Cornblath and Hartmann, 1948; Jaffe, 1981; Marshall et al., 1945; Knotek and Schmidt, 1964; WHO, 1977a). Recently, methaemoglobinemia has been reported in all age groups (Gupta et al., 2000). It was observed that methaemoglobinemia was prevalent among all age groups, but was more severe in infants and older age groups (>45 years). Once methemoglobin levels in blood exceeds 10 per cent of total Haemoglobin, it manifests as clinical cyanosis and causes cellular anoxia. Effect of methemoglobin (in relation to percentage) and clinical presentation are given below (WHO, 1977c):

Meth-Hb	*Clinical presentation*
<10%	No signs and symptoms
10–25%	No symptoms
	Cyanosis Present
25–50%	Cyanosis, Dyspnoea, Headache
50–60%	Dyspnoea even on lying, Cyanosis, Disorientation
60%	Lethal levels

In several Indian villages people have been consuming water containing high nitrate concentration, at times up to 1,200 mg nitrate ion/L. High nitrate concentration are causing methaemoglobinemia in all ages (Gupta et al., 2000). It was observed that this methaemoglobinemia was more in infants and elderly people (ibid.). Children and adolescents have lower levels of methemoglobin. Adaptation to the reserve of cytochrome b_5 reductase activity with increasing water nitrate concentration to compensate methaemoglobinemia was reported to be a possible mechanism. Gupta et al., (1999b) indicated that high nitrate concentration is hazardous not only to infants but also to groups over 18 years of age. It was reported that adaptation of cytochrome b_5 reductase activity peaks at about 95 mg nitrate ion/l nitrate concentration and gets exhausted by nitrate level 200 mg nitrate ion/l, hence making people more prone for toxicity.

Infant Mortality Rate (IMR)

A study (Super et al., 1981) on African mothers and other studies also (Schwiede et al., 2005; Fewtrell, 2004; Spalding and Exner, 1993; CDC, 1996; Dorsch et al., 1984) reported an increase in infant deaths with increasing exposure of pregnant mothers and infants to nitrate. This may either be due to undetected toxic methaemoglobinemia or to malformations and weaknesses in the infant caused by foetal nitrate exposure. It was further suggested that because of high IMR there is need to revise the nitrates standards for drinking water (Kumar et al., 2002).

Nitrate, Nitrite, Nitrosamines and Cancer

Nitrate itself is not carcinogenic, but instead acts as a 'procarcinogen', meaning that it reacts with other chemicals (amines and amides) to form carcinogenic compounds (N-nitroso compound). Under acidic conditions, nitrites (NO^{2-}) are formed naturally from nitrates, and nitrites in turn may combine with amines or amides to form N-Nitroso compounds (nitrosamines). The endogenous formation of carcinogenic N-nitroso compounds can occur following ingestion of nitrate from drinking water. Nitrate is first reduced in the saliva to nitrite, which can react in the stomach with secondary amines and amides. The nitrosamines most commonly reported in foods are dimethylnitrosamine, diethylnitrosamine, nitrosoproline and nitrosopyrrolidine. Nitrosamines can also be found in some alcoholic beverages and in tobacco products. The amount of N-nitroso compounds which can be formed in vivo depends in part on the availability of nitrite, which is itself dependent on the availability of nitrate, the presence of a microbial population with nitrate reductase activity and conditions favourable to chemical nitrosation (Tannenbaum, 1983). The physiological studies (Weisenberg et al., 1982; NAS, 1977; Moller et al., 1989) provide strong support, indicating the association between nitrate contamination of drinking water and increased cancer rates. It has been reported that endogenously formed N-nitroso compounds are important in human cancer (Michaud, 2004; Mirvish et al., 1995). Szaleczky et al. (2000) reported that endogenously formed nitrogen and oxygen free radicals are believed to be involved in human cancer etiology. They reported that plasma nitrate/nitrite originates from endogenous nitric oxide production in fasting humans, decrease in superoxide scavenger activity (SSA) and free sulfhydryl groups (SH) reflects the amount of superoxide anion generated and nitrotyrosine is believed to be formed by the interaction of tyrosine and peroxynitrite.

Population ingesting larger amounts of nitrate might be expected to have a higher incidence of cancer of the relevant target organ (NAS, 1981). In addition, products of nitric oxide, generated by macrophages during inflammation, can react with water at neutral pH to form nitrite and nitrate and with amines to form nitrosamines (Mirvish, 1995). Hence, the International Agency for Research on Cancer (IARC, 1978; Fraser et al., 1980) has classified nitrate/nitrite as possibly carcinogenic to humans.

N-Nitroso compounds have been associated with 15 different types of cancers, including tumours in the bladder, stomach, brain, esophagus, bone and skin, kidney, liver, lung, oral and nasal cavities, pancreas, peripheral nervous system, thyroid, trachea, acute myelocytic leukemia and T and B cell lymphoma. This group found to be carcinogenic in a wider range of tumours than any other group of carcinogens (NAS 1977, 1978; IARC 1978; Mirvish, 1991 and 1983). More than one hundred of these N-Nitroso compounds have been tested for carcinogenicity in animals and 75–80 per cent of them have been found to be carcinogens (NAS, 1977). IARC (1978) concluded that 11 N-Nitroso compounds were carcinogenic in man. In humans, the organs thought to be most at risk from cancers are the stomach, esophagus, nasopharyngeal cancer, and cancer of the bladder.

To summarize, the carcinogenesis due to high nitrate ingestion is multifactorial. The possible mechanism may be:

1. Increased formation of nitrosamines (known carcinogen), a metabolic intermediate product of nitrate metabolism,
2. Increased formation of free oxide radicals,
3. Interference in NO metabolism.

Still there are studies (RCEP, 1979; Fraser et al., 1980; Zaldivar, 1977; Armijo and Coulson, 1975; Cuello et al., 1976; Hill et al., 1973) which have drawn no firm conclusion to prove nitrate as carcinogen.

Nitrate, Nitrite, Nitrosamines and Cancer in Infants and Children

A series of human and animal studies have indicated that infants, children and even the foetus may face elevated risk later in life due to the effects of nitrate or nitrite exposure. Human epidemiology studies found an increase in stomach cancer rates with consumption of well water high in nitrate and that individuals exposed during the first ten years of life formed a high risk group (Gray et al., 1991; Cuello et al., 1976). Studies with N-Nitroso compounds on both foetal and infant equivalent animals support this finding (Gray et al., 1991; Cuello et al., 1976). Animal studies have documented transplacental passage of nitrite and reported that a high dose of nitrate to the pregnant ham can cause subacute methaemoglobinemia in foetal rats (Shuval and Gruner, 1972). Further animal studies with nitrosamine compounds showed that exposure during infancy increases the cancer risk from N-nitroso compounds by a factor of six (Gray et al., 1991).

Nitrate and Gastric Cancer

Since 1976, a relationship between increasing rates of stomach cancer and increasing nitrate intake have been documented (Mirvish, 1983; Armijo, 1975, 1981; Cuello et al., 1976; Xu and Reed, 1992; Reed et al., 1981, 1983). An increased risk of gastric cancer in conditions associated with low gastric acidity is well recognized and lends support to the hypothesis that N-nitroso compounds may be involved in its development. High levels of gastric juice nitrites and elevated urine levels of nitrosamines have been reported in patients with chronic atrophic gastritis or high intragastric pH levels (Farinati et al., 1996; Jaskiewicz et al., 1990; Farinati et al., 1989; Vermeer et al., 2001; Dallinga et al., 1998). High nitrate intake with low Vitamin C intake, high meat intake, or chronic bowel inflammation has been shown to be associated with increased cancer incidence. A study suggests that the *Cyp2E1 RsaI* polymorphism is associated with a reduced risk of gastric cancer (Nishimoto et al., 2000).

Reed et al. (1983) reported first time in humans a significant lowering of gastric juice and N-nitroso compounds by ascorbate treatment in 51 achlorhydric subjects. This may be an important observation for preventing gastric cancer in high-risk subjects (NAS, 1981; WHO, 1985b, 1996; ECETOC, 1988; RIVM, 1989).

Nitrate and Non-Hodgkins Lymphoma (NHL)

Epidemiological study of nitrate in well water in Nebraska showed an association between nitrate contamination and a different kind of cancer, non-Hodgkins lymphoma (Weisenburger, 1990). The author concluded that 'NHL in eastern Nebraska may be related to the use of pesticides and nitrogen fertilizers'.

Nitrate and Colorectal Cancer

De Roos et al., (2003) observed negligible overall associations of either cancer with average nitrate level and with the number of years with average nitrate-nitrogen level greater than 5 or 10 mg/l. They reported an increased risk of colon cancer associated with drinking water nitrate among certain subgroups expected to have high rates of nitrosation, but they did not observe the same patterns for rectum cancer. The colon and rectum have similar epithelial tissues, but these two cancer types have somewhat different risk factors. Although N-nitroso compounds formed in the digestive tract would be expected to pass through the rectum, contractile activities in the rectum cause faecal matter to pass through quickly, resulting in less contact time with the rectum than with the colon (Vander et al., 1994).

Nitrate and Urinary Bladder Cancer

Recently, studies have pointed out risk of bladder cancer and high nitrate ingestion. Nitrate ingestion contributes to stomach nitrite levels, which can react with amines (nitrosation) to form endogenous nitrosamines (Mirvish, 1995). In a recent epidemiological study, drinking water nitrate levels were associated with a significant elevation in bladder cancer risk (Weyer et al., 2001). In addition, products of nitric oxide, generated by macrophages during inflammation, can react with water at neutral pH to form nitrite and nitrate and with amines to form nitrosamines (Mirvish, 1995). High levels of gastric juice nitrites and elevated urine levels of nitrosamines have been reported in patients with chronic atrophic gastritis or high intragastric pH levels (Farinati et al., 1996, 1989; Jaskiewicz et al., 1990; Vermeer et al., 2001). Significant correlations between mean intragastric pH values and mean *N*-nitrosodi-*n*-butylamine level (P=0.005) and total volatile *N*-nitrosamine contents (P=0.009) were also observed (Dallinga, 1998).

Respiratory system

A correlation among drinking water nitrate concentration, high methemoglobin levels and pathological changes in bronchi and lung parenchyma have been reported in animal studies (Gruener and Shuval, 1970). Changes in lungs reported were frequent dilation of bronchi with lymphocytic infiltration of mucosa and muscles, frequent purulent bronchial exudates, interstitial round cell infiltration and fibrosis at certain areas. WHO (1977d) reported an association of increased asthmatic attacks and high air borne nitrate concentration.

A high percentage (40–82%) of cases of acute respiratory tract infection with history of recurrence has been reported in children drinking high nitrate in water (Gupta et al., 2000). These findings were further substantiated (Gupta et al., 1999c) in an animal experiment on rabbits. Significant changes in lungs were observed with congestion, presence of inflammatory cells and breakdown of alveoli. These changes were absent when animals were fed water containing 45 mg/l of nitrate, as the concentration of nitrate increased the changes were more pronounced and severe.

Nox, Tobacco and Malignancy

Tobacco use causes 20 per cent of cancer deaths worldwide. The International Agency for Research on Cancer predicts 10 million tobacco-related deaths annually by 2020, of which 70 per cent will occur in the developing world (IARC, 2004)

Study conducted by Sleiman et al. (2010) reported that residual nicotine from tobacco smoke sorbed to indoor surfaces reacts with ambient nitrous acid (HONO) to form carcinogenic tobacco-specific nitrosamines (TSNAs). Substantial levels of TSNAs were measured on surfaces inside a smoker's vehicle.

Time-course measurements revealed fast TSNA formation, with up to 0.4 per cent conversion of nicotine. Given the rapid sorption and persistence of high levels of nicotine on indoor surfaces – including clothing and human skin – this recently identified process represents an unappreciated health hazard through dermal exposure, dust inhalation and ingestion.

These findings raise concerns about exposures to the tobacco smoke residue that has been recently dubbed – third hand smoke (SHS, smoke inhaled unintentionally).

The work of Sleiman et al. (2010) highlighted the importance of reactions at indoor interfaces, particularly those involving amines and NOx/HONO cycling, with potential health impacts. Whereas direct inhalation of SHS is an exposure pathway of concern, non-smokers, especially infants, are at risk through contact with surfaces and dust contaminated with residual smoke gases and particles (Matt et al., 2004). Reactions of atmospheric species [O_3, nitrous acid (HONO), NOx] with residual smoke on surfaces (furniture, walls, skin and clothing) have been overlooked as a source of long-term exposure to harmful pollutants. Nicotine, their precursor, is the most abundant organic compound emitted during smoking (up to 8 mg per cigarette). It is deposited almost entirely on indoor surfaces and persists for weeks to months (Singer et al., 2003, 2004). HONO is often present in indoor environments at higher levels than outdoors.

Typical indoor levels are 5–15 ppbv, with (HONO)/(NO2) ratios ~0.15 to 0.4 (vs.~0.03 outdoors). Indoor levels up to 100 ppbv have been reported (Finlayson-Pitts and Pitts, 2000; Wainman et al., 2001; Lee, et al., 2002). The main indoor sources of HONO are direct emissions from unvented combustion appliances (Spengler, 1993; Pitts et al., 1989), smoking (Eatough et al., 1989) and surface conversion of NO_2 and NO (Goodman et al., 1999; Finlayson Pitts et al., 2003; Mertes and Wahner, 1995; Hirokawa et al., 2008; Yabushita et al., 2009; Enami et al., 2009). Heterogeneous formation of HONO also occurs inside automobiles, leading to (HONO) up to 30 ppbv and (HONO)/(NO2) ~ 0.4 in polluted urban areas (Febo and Perrino, 1995).

Pitts et al. (1978) first described the atmospheric production of N-nitrosamines by reactions of nitrogen oxides and HONO with amines. N-nitrosamines were found to be unstable in sunlight, rendering the reaction unimportant in outdoor daytime conditions. However, this process can be relevant in indoors where N-nitrosamines and HONO are less vulnerable to photochemical decomposition.

Cardiovascular System

The cardiac toxicity of nitrate is mediated through its metabolic effects and metabolic products, e.g. metheoglobin formation, increased NO formation, increased in free oxide radicles (ROS) and high levels of compensatory enzymes (Cytochrome b_5 reductase, etc.). The following effects have been reported:

1. Early onset of hypertension
2. Inflammation and degeneration of cardiac musculature
3. Causing endothelial dysfunction
4. Making the person prone for ischaemia
5. Enhances arteriosclerotic and atherosclerotic process

Early Onset of Hypertension

Earlier onset of hypertension has been reported with high nitrate ingestion (Malberg, 1978). Some studies also reported an inverse relationship between cardiovascular mortality and nitrate concentration in water supplies (Morton, 1971). Pomeranz et al., (2000) reported that elevated salt and nitrate levels in drinking water cause an increase of blood pressure in school children.

Inflammation and Degeneration of Cardiac Musculature

The nitrate toxicity on cardiovascular system has been reported in both animals and human beings. Animal studies in early seventies (Shuval and Gruener, 1972; Gruener and Shuval, 1970) reported correlation among drinking water nitrate concentration, high methemoglobin levels and cardiac muscles. The changes reported in cardiac muscles were small foci of inflammatory cells and fibrosis. Diffused interstitial cellularity with pronounced degenerative foci was frequent in the highest nitrate groups only. The intramural coronary arteries provided surprising results with some degree of thinning and dilation in comparison to the control group who showed some degree of thickening, a marked hypertrophy and narrowing. Further, Gupta et al. (1999a) in an animal experiment reported that high nitrate ingestion was associated with changes in cardiac muscles in the form of branching of myosites, presence of inflammatory cells and focal degenerative changes in cardiac muscles. These changes were absent when animals were fed with water containing 45 mg/l of nitrate, but as the concentration of nitrate increased the changes were more pronounced and severe.

The changes in the cardiac muscles with different concentration of nitrate in water are given below (Table 9.2).

Table 9.2

The Effect of Nitrate in Drinking Water on Cardiac Muscles.

Water Nitrate (mg/l)	*Branching of Myosites*	*Inflammatory Cells*	*Focal Degenerative Changes*
45	NIL	NIL	NIL
100	+	NIL	NIL
200	+	NIL	NIL
400	+	+	+
500	+	+	+

(+) Mild, (++) Moderate, (+++) Severe, NIL: Normal histopathology (No change observed)

The changes in cardiac tissue even in animal study may be of importance, in view of the side/adverse effects related to the use of nitrate containing drugs for the management of cardiac disorders and increasing drug tolerance of these nitrate containing drugs (USP DI, 1990).

Endothelial Dysfunction

Endothelial dysfunction, or the loss of proper endothelial function, is a hallmark for vascular diseases and is often regarded as a key early event in the development of atherosclerosis. Impaired endothelial function is often seen in patients with coronary artery disease, diabetes mellitus, hypertension, hypercholesterolemia, as well as in smokers. Endothelial dysfunction has also been shown to be predictive of future adverse cardiovascular events. One of the main mechanisms of endothelial dysfunction is the diminishing of nitric oxide, often due to high levels of asymmetric dimethylarginine, which interfere with the normal L-arginine-stimulated nitric oxide synthesis. The most prevailing mechanism of endothelial dysfunction is an increase in reactive oxygen species (ROS), which can impair nitric oxide production and activity via several mechanisms. (Toledo et al., 2004; Deanfield et al., 2005) The signaling protein ERK5 is essential for maintaining normal endothelial cell function (Roberts et al., 2009).

An association has been observed between increase in plasma nitrite/nitrate levels and an excessive production and/or inactivation of NO leading to impaired vascular smooth-muscle reactivity leading to impairment in endothelium dependent, shear stress-induced vasodilatation.

Making the Person Prone for Ischemia

Nitrate and nitrite have emerged as viable alternative source of NO under ischemic conditions. NO has been shown to be one of the most important molecules for the prevention of injury from Ischemia. Paradoxically, in conditions of inadequate oxygen, the NO is dangerous and may be the cause of Ischemia (Garg and Bryan, 2009).

Nitrate inhalation exposure for long term, of individuals working in munitions or dynamite manufacturing leads to an apparent biocompensation of the cardiovascular effects of organic nitrates. These workers, when removed from the source of the nitrates, experienced symptoms of angina and were subject to sudden and sometimes fatal heart attacks (Carmichael and Lieben, 1963; Stokinger, 1982).

Atheroscerosis

Nitrate is a source of ROS in human body. ROS is known to cause enhanced atherosclerosis process. Atherosclerosis develops from low-density lipoprotein molecules (LDL) becoming oxidized (ox-LDL) by free radicals, particularly reactive oxygen species. When oxidized LDL comes in contact with an artery wall, a series of reactions occur to repair the damage to the artery wall caused by oxidized LDL. The LDL molecule is globular shaped with a hollow core to carry cholesterol throughout the body. Cholesterol can move in the bloodstream only by being transported by lipoproteins (Anthea et al., 1993). The body's immune system responds to the damage of artery wall caused by oxidized LDL via sending specialized white blood cells (macrophages and T-lymphocytes) to absorb the ox-LDL and forming specialized foam cells.

Unfortunately, these white blood cells are not able to process the ox-LDL and ultimately grow then rupture, depositing a greater amount of oxidized cholesterol into the artery wall. This triggers more white blood cells to make a vicious cycle (Sloop and Garber, 1997). Eventually, the artery becomes inflamed. The cholesterol plaque causes the muscle cells to enlarge and form a hard cover over the affected area. This hard cover is what causes a narrowing of the artery, reduces the blood flow and increases blood pressure (Duguid, 1960).

It is important to discuss the preventive role of thioredoxin proteins to counteract the toxic effects of nitrate. Zhang et al. (2007) demonstrate regulation of such events by a mitochondria-specific thioredoxin, which reduces oxidative stress and increases NO bioavailability, thus preserving vascular endothelial cell function and preventing atherosclerosis development. Thioredoxin proteins are classically defined by their ability to reduce disulfides (a SOS bond) to dithiols (two OSH groups) and in the process the thioredoxins are oxidized from a dithiol to a disulfide. The thioredoxin disulfide is then cycled back to its active form by the enzyme thioredoxin reductase, using NADPH as the electron donor (Mustacich and Powis, 2000; Arnér and Holmgren, 2000). The role of thioredoxin proteins as an antioxidant that reduces protein thiols and various forms of ROS is widely appreciated.

It has long been described that superoxide rapidly reacts with NO to inactivate it. In fact, this was a major clue to the original identification of endothelial-derived relaxing factor as NO. The Superoxide reacts directly with NO with high affinity to form the reactive nitrogen species peroxynitrate. In addition to inactivating NO, ROS have a host of potentially deleterious effects and antioxidants are widely touted as having some potential for heart disease prevention. However, initial results from clinical trials of antioxidants have been disappointing, suggesting that we have insufficient understanding of the specific roles of ROS and antioxidants to design effective therapies.

Some studies reported that inorganic nitrate from dietary and endogenous sources are emerging as a substrate for in vivo generation of nitric oxide (NO) and other reactive nitrogen oxides. Although blood pressure reduction, inhibition of platelet aggregation and vasoprotective activity have been also reported. We may have to reconsider our current thinking and realize that inorganic nitrate may not necessarily be a threat to human health. Instead, in some years, we might even consider this anion as an essential nutrient. But still it is a long way to go.

Gastrointestinal System

In gastrointestinal system three major findings have been reported:

1. Recurrent diarrhoea in children,
2. GI Malignancy,
3. Histopathological changes in animal study.

Recurrent Diarrhoea in Children

Gupta et al. (2001) reported a problem of recurrent diarrhoea in children up to 8 years of age. They suggested that increased consumption of nitrate leads to (a) increased production of nitrite (Allison and Reddy, 1984; Cole and Brown, 1980); (b) enhanced absorption of sodium from the intestinal lumen (Roediger et al., 1986); (c) excess NO (free radical nitric oxide) generation, having vasodilatory effect (Winterbourn et al., 1976; Hsia, 1998; Berger et al., 1997; Smith et al., 1997); and (d) increased production of O_2^-, which will react with other cell constituents, possibly causing irreversible cell damage (Berger et al., 1997; Gupta et al., 1998). These changes in enteric mucosa cause hyperemia and edema in the enteric mucosa and later on, possibly cause irreversible mucosal damage and therefore, provide high-risk conditions suitable for recurrent diarrhea.

These findings were correlated well with histopathological study conducted on rabbits (Gupta et al., 2001). It was observed that the histopathological changes observed in the intestinal mucosa on rabbit study revealed that the degree of damage in colon was progressive as the nitrate content of the ingested water increased.

These findings are of interest since infants and children consuming nitrate rich water, especially during diarrhoea which is an aggravating factor for nitrate toxicity (Murray and Chirstie, 1993), where the use of WHO oral rehydration solution (ORS) is a normal routine. The use of WHO ORS could be of grave concern, if prepared with water containing high nitrate.

GI Malignancy

The formation of N-nitroso compounds in the stomach has been connected with drinking water nitrate and excretion of N-nitroso compounds by humans has been associated with nitrate intake at the acceptable daily intake level through drinking water (Vermeer et al., 1998). The metabolism of nitrate and nitrite, the formation of N-nitroso compounds and the development of cancers in the digestive system are complex processes mediated by several factors. Individuals with increased rates of endogenous formation of carcinogenic N-nitroso compounds are likely to be susceptible. Known factors altering susceptibility to the development of cancers in the digestive system are inflammatory bowel diseases, high red meat consumption, amine-rich diets, smoking and dietary intake of inhibitors of endogenous nitrosation (e.g. polyphenols and Vitamin C) (Vermeer et al., 1998; De Kok et al., 2005; De Roos et al., 2003). In 1995, when the Subcommittee on Nitrate and Nitrite in Drinking Water reported that the evidence to link nitrate to gastric cancer was rather weak (NRC, 1995), the stomach was still thought to be the most relevant site for endogenous nitrosation. Previous studies, such as those reviewed in the NRC (1995) report, which found no link between nitrate and stomach cancer, concentrated on the formation of nitrosamines in the stomach. Recent work indicated that larger amounts of N-nitroso compounds can be formed in the large intestine (De Kok et al., 2005; Cross et al., 2003).

Since 20–90 per cent of flatus is formed by Nitrogen, further ingestion of high nitrate water or constipation enhances the nitrate toxicity. The major components of the flatus, which are odourless, by percentage are:

- *Nitrogen:* 20–90 per cent
- *Hydrogen:* 0–50 per cent
- *Carbon dioxide:* 10–30 per cent
- *Oxygen:* 0–10 per cent
- *Methane:* 0–10 per cent

Histopathological Changes in Animal Study

Animal studies conducted by Gupta et al. (2001) indicated changes in structure of liver and intestine with high nitrate ingestion. Further, it was observed that the pathological changes were more with increasing nitrate concentration. The changes observed were:

Liver: Hepatocytes are more eosinophilic and grannular. Hepatic cords are distorted. Hepatocytes show marked degeneration. Marked congestion was noted. Inflammatory cells were observed. Degenerative changes of nucleus were found. Portal tract shows slight congestion and presence of inflammatory cells. These changes were more pronounced as the nitrate concentration in water increases.

Intestine: The histopathological changes in the intestine were not observed at 100 ppm nitrate but as the nitrate concentration in drinking water increases from 100 ppm, focal collection of inflammatory cells starts at sub mucosa with slight lymphoid hyperplasia, which was more pronounced as the water nitrate concentration reaches to 400 ppm and at this level degenerative changes also starts.

Abortions

Health effects associated with ingestion of nitrate-contaminated water have included stillbirth, low birth weight and slow weight gain and even death of the animals affected (Committee on Nitrate Accumulation, 1972). Spontaneous abortions were also observed in laboratory animals and livestock (FDA, 1972; Sund et al., 1957). In 1959, for the first time in humans, spontaneous abortions were reported to have an association with increased methemoglobin levels due to high nitrate ingestion (Muhrer et al., 1959). In subsequent years some studies (Schmitz, 1961; CDC, 1996; Fewtrell, 2004) found an increased risk of spontaneous abortion or certain birth defects if the mother drank water high in nitrate. Therefore, it has been suggested that women who are pregnant or who are trying to become pregnant should not consume water containing high nitrate.

Birth Defects – Malformations

The risk of birth defects due to nitrate exposure is a particular concern because of the fact that risk could be due to a single high dose of nitrate early in the pregnancy that later has profound effects on long-term foetal development. Animal studies have indicated that there is transplacental transfer of N-Nitroso compounds to the foetus (Shuval and Gruener, 1972) and this foetal exposure can cause cancer later in life (Druckrey et al., 1966). It was also reported that a single dose of a nitrosamide given to pregnant rats on day 15 of the pregnancy was reported to cause birth defects in the offspring. In rats and hamsters studies multiple birth defects, including malformations of the eye, central nervous system and musculoskeletal system, were observed when a single dose of N-Ethyl-N-Nitrosourea, a nitrosamine, was given to the mother (Druckery et al., 1966; Givelber and Di Paolo, 1969).

Extending these finding on human beings, a number of human epidemiology studies were conducted (Dorsch et al., 1984; Knox, 1972; Super, 1981). A link was found between anencephaly rates and intake of cured meat containing high levels of nitrite (Knox, 1972). This study provided the first suggestive evidence in humans that nitrite consumption in food could have adverse impacts on the foetus. A study (Dorsch et al., 1984) found statistically significant dose response relationships between birth defects of the central nervous system and musculoskeletal system and increasing nitrate concentration of drinking water. Arbuckle et al. (1988) found the evidence for an association between nitrate and birth defects to be weaker. Fan (1987) reported that nitrate contamination of 45 ppm nitrate ion, or 10 ppm nitrate-nitrogen, adequately protects the very young from nitrate-induced toxicity, both pre- and post-natal. The adverse reproductive effects reported occurred at doses that were about one thousand times and higher than the estimated human intake. Croen et al. (2001) found that exposure to nitrate in groundwater at concentration above the 45 mg/l maximum contaminant level was associated with increased risk for anencephaly (Odds ratio 4.0; confidence interval 1.0–15.4). Risk for anencephaly increased for mothers with the highest nitrate exposure 36–67 mg/l in groundwater, compared with nitrate exposure less than 5 mg/l (Odds ratio 6.9; confidence interval 1.9–24.9).

Diabetes

A positive correlation between high nitrate levels in drinking water and increased incidence of Type 1 diabetes was observed independently from length of mother's education, child's or mother's age, place of residence or mother's smoking status (Kostraba 1992; Virtanen et al., 1994; Parslow et al., 1997; Van Maanen et al., 1994). Further studies indicated that consumption of high levels of N-nitroso compounds (NOCs) by human mothers may result in an increased incidence of Type 1 diabetes in male offspring (Helgason and Jonasson, 1981). In animal studies NOCs have been reported to be toxic to pancreatic beta cells, providing the rationale for these observations (Wilson et al., 1983). Kostraba (1992) postulated that exposure to nitrate in drinking water causes increased production of free radicals, which may play a role in the etiopathogenesis of insulin-dependent diabetes mellitus (IDDM). The production of free oxide radicals following high nitrate ingestion have been further supported by studies conducted by Gupta et al. (1999a and 2000).

Further it is to be noted that there are studies, which indicated no relationship between nitrites and nitrates in drinking water and increased incidence of Type 1 diabetes (Dahlquist et al., 1990; Virtanen et al., 1994; Verge et al., 1994). Thus, while some studies suggest an association between intake of nitrate/nitrites and risk of Type 1 diabetes, the data remain limited and inconsistent (Longnecker and Daniels, 2001). Hence, the association of dietary nitrate/nitrites with diabetes remains tenuous and further research needs to be supported.

IDDM is an autoimmune disease characterized by the selective destruction of insulin secreting pancreatic â-cells found in the islets of Langerhans (Gepts, 1965). Since nitrates are source of nitric oxide, it has been shown that nitric oxide, produced in micromolar levels by the â-cell in response to cytokines, mediates the damaging effects of cytokines through the inhibition of mitochondrial enzymes, inhibition of glucose stimulated insulin secretion and the induction of DNA strand breaks and base modifications (Hughes et al., 2009; Corbett et al., 1992a; Southern and Green, 1990).

Cytokines, released from invading leukocytes during insulitis, are believed to participate in the initial destruction of â-cells and precipitating in the autoimmune response (Mandrup-Poulsen, 1996; Rabinovitch et al., 1998).

Nitric oxide, produced in micromolar levels following enhanced expression of the inducible nitric oxide synthase (iNOS) in â-cells, mediates the damaging actions of cytokines on â-cell function (Corbett et al., 1992a, 1992b; Southern and Green, 1990). Nitric oxide inhibits insulin secretion by attenuating the oxidation of glucose to CO_2, reducing cellular levels of ATP and, thereby, attenuating ATP-inhibited K^+-channel activity (Misler et al., 1992; Hughes et al., 1990). The net effect is the inhibition of â-cell depolarization, calcium entry and calcium-dependent exocytosis. In addition to the inhibition of â-cell function, nitric oxide induces DNA damage in â-cells (Eizirik et al., 1996; Steer et al., 2006). Nitric oxide, or the oxidation products N_2O_3 and $ONOO^-$, induce DNA damage through direct strand breaks and base modification (Burney et al., 1999; Tamir et al., 1996; Niles et al., 2006) and also by inhibition of DNA repair enzymes, thereby enhancing the damaging actions of nitric oxide (Jaiswal et al., 2001; Graziewicz et al., 1996). Recent studies have shown that â-cells maintain a limited ability to recover from cytokine-mediated damage (Scarim et al., 1997; Corbett 1992a).

Thyroid

Nitrate competitively inhibits iodine uptake. If dietary iodine is available at an adequate range (corresponding to a daily iodine excretion of 150–300 μg/day), the effect of nitrate is weak, with a tendency to zero. The nitrate effect on thyroid function is strong if a nutritional iodine deficiency exists simultaneously (Höring et al., 1991; Höring H, 1992). An association between high nitrate concentration in drinking-water and goiter has been (Höring et al., 1991; Höring, 1992; Van Maanen et al., 1994) described by various studies. A dose-response relationship has been demonstrated by Höring H. et al. (1991) (nitrate in drinking-water v/s incidence of goiter) as well as by Van Maanen et al. (1994) (nitrate in drinking-water vs. thyroid volume). Both the experimental and epidemiological studies give the impression that nitrate in drinking water has a stronger effect on thyroid function than nitrate in food.

The nitrate is found in a high concentration in soil, agricultural products and drinking water, because of the use of fertilizer. The people living in high nitrate polluted area have a chronic nitrate exposure.

It has been shown that the children exposed to water pollution by nitrates have a higher relative risk of goiter (Vladeva et al., 2000). It has been reported that the thyroid gland volume of the school children and adults living in high nitrate polluted area is higher than those living in low nitrate polluted area (Van Maanen et al., 1994; Tajtakova et al., 2006). It has also been reported that the frequency of hypoechogenicity in school children living in high nitrate polluted area was higher than those living in low nitrate polluted area (Tajtakova et al., 2006). Various experimental studies showed that especially long-term nitrate exposure via drinking water may result in an increase in thyroid gland weight (Eskiocak, 1995; Eskiocak et al., 2005; Zaki et al., 2004) and histomorphological changes, including retention of lobular architecture, prominent vascular congestion, follicular hyperplasia, a vacuolisation and an increase in the colloidal volume of the follicles thyroid (Eskiocak et al., 2005; Zaki et al., 2004) Several investigators showed that low dose or short-term nitrate intake causes a decrease in thyroid radioiodine uptake (Katti and Sathyanesan, 1987; Shrivastava et al., 1987; Szokeova et al., 2001).

Several research groups reported that nitrate causes thyroid hormone dysfunction. It found that the thyroid secretion rate is significantly reduced in the goat which intake 50 mg/l nitrate via water in comparison with the control group (Simon et al., 2000). Van Maanen et al. (1994) reported that there is a decrease in TSH levels in humans exposed to high nitrate pollution, whereas Tajtakova et al. (2006) reported that an increase in TSH levels in schoolchildren living in high nitrate pollution area. There are also some experimental studies which support that thyroid hormone dysfunctions may be caused by nitrate intake via drinking water. It has been reported that the total and free T_3 and T_4 levels are decreased at high dose and long-term nitrate exposure (Eskiocak, 1995; Eskiocak et al., 2005; Jahreis et al., 1987). However, Szokeova et al. (2001) demonstrated that short-term nitrate administration may result in a significantly higher serum level of total T_3. These findings indicate that short- or long-term nitrate exposure may be strongly responsible for the prominent change in thyroid hormone production.

Several investigators showed that low dose or short-term nitrate intake causes a decrease in thyroid radioiodine uptake (Katti and Sathyanesan, 1987; Shrivastava et al., 1987; Szokeova et al., 2001). Whereas studies (Eskiocak, 1995; Eskiocak et al., 2005) reported that high dose and long-term nitrate exposure result in an increase in the thyroid radioiodine uptake. These findings suggest that the effect of nitrate on thyroid iodine uptake is dose-dependent and the inhibition of thyroid iodine uptake may be stronger at higher amounts of nitrate.

Several research groups reported that nitrate cause thyroid hormone dysfunction. It found that the thyroid secretion rate is significantly reduced in the goat which intake 50 mg/l nitrate via water in comparison with the control group (Simon et al., 2000). Van Maanen et al. (1994) reported that there is a decrease in TSH levels in humans exposed to high nitrate pollution, whereas Tajtakova et al. (2006) reported that an increase in TSH levels in schoolchildren living in high nitrate pollution area. There are also some experimental studies which support that thyroid hormone dysfunctions may be caused by nitrate intake via drinking water. It has been reported that the total and free T_3 and T_4 levels are decreased at high dose and long-term nitrate exposure (Eskiocak, 1995; Eskiocak et al., 2005; Jahreis et al., 1987). However, Szokeova et al. (2001) demonstrated that short-term nitrate administration may result in a significantly higher serum level of total T_3. These findings indicate that short- or long-term nitrate exposure may be strongly responsible for the prominent change in thyroid hormone production.

Adrenal Gland

High nitrate ingestion affects the adrenal gland. In addition to the effect of nitrite on the adrenal zona glomerulosa in rats, a study in humans indicated that sodium nitrite (0.5 mg of sodium nitrite per kg of body weight per day, during 9 days) causes a decreased production of adrenal steroids, as reflected by the decreased concentration of 17-hydroxysteroid and 17-ketosteroids in urine (Til et al., 1988; Kuper and Til, 1995).

Similar results were also found in rabbits (Violante et al., 1973). Although the mechanism is not clear, the effects of nitrite seen in rats seem relevant for the hazard assessment in humans.

Immunity

Few studies (Ustyugova et al., 2002; Kozliuk et al., 1989) had reported the effect of nitrate/nitrite ingestion on immune system. The effect of nitrate ingestion (Ustyugova et al., 2002) on human immune system indicated that nitrate had no effect on lymphocyte growth, but nitrite decreases proliferation of lymphocytes. Fibroblast growth remains unaffected. A decreased production of Th1 cytokines (interleukin–2, interferon-gamma and tumour necrosis factor-beta), which is responsible for resistance to a variety of infectious diseases was noted. No effect on the production of the Th2 cytokine interleukin–10, which is responsible for disease susceptibility, was noted. Because nitrate/nitrite shifted the balance from a Th1 to a Th2 response in some individuals, exposure to these compounds may decrease these people's responsiveness to infectious diseases. The levels of nitrate used in this study are relevant to human health because they are present in the liquid portion (non-breastfed) of some 2-month-old infant diets in rural Romania. Animal studies also reported an immune suppression due to high nitrate ingestion (Porter et al., 1999).

Air Pollution and Nitrate Toxicity

Nitrogen oxides or NOx is the generic term for a group of seven compounds, highly reactive gases, all of which contain nitrogen and oxygen in varying amounts. The most important of these are nitric oxide (NO), nitrogen dioxide (NO_2) and nitrous oxide (N_2O). Nitric oxide and nitrogen dioxide are air pollutants. Nitrous oxide is not generally considered to be an air pollutant, but it is an ozone depleting gas and also contributes significantly to visibility impairment in the form of plumes and hazes. Most of the nitrogen oxides in our environment come from natural rather than anthropogenic sources. Natural sources of NOx are lightning and biological and non-biological processes in soil; they also descend from the stratosphere.

NOx cause a wide variety of health and environmental impacts because of various compounds and derivatives in the family of nitrogen oxides, including nitrogen dioxide, nitric acid, nitrous oxide, nitrates and nitric oxide. They react to form nitrate particles, acid aerosols and contribute to formation of acid rain. NOx constitute one of the main ingredients involved in the formation of ground level ozone, which can cause serious respiratory problems because of its potent oxidizing properties. This occurs when NOx and volatile organic compounds (VOCs) react in the presence of heat and sunlight.

Nitrogen oxides form when fuel is burned at high temperatures, as in combustion in power plants and automobiles and during processes used in chemical plants. The general population is primarily exposed to nitrogen oxides by breathing in air polluted with these gases. In addition, people residing in areas close to coal burning power plants or areas with heavy vehicular traffic are mainly exposed to higher levels of nitrogen oxides. Households that burn biomass fuels or coal or gas stoves have higher levels of nitrogen oxides in them when compared to houses using electrical appliances. Another source of indoor NOx is tobacco smoke. Active or passive inhalation of second-hand smoke exposes one to nitrogen oxides. Occupationally, workers employed in facilities that produce nitric acid or certain explosives like dynamite and trinitrotoluene (TNT), as well as workers involved in the welding of metals may breathe in nitrogen oxides during their work.

Health effects are related to levels of NOx as well as the duration of exposure (Bernstein et al., 2004). Low levels of nitrogen oxides in the air irritate eyes, nose, throat and lungs leading to cough and shortness of breath, tiredness and nausea. Breathing high levels of nitrogen oxides can cause rapid burning, spasms and inflammatory swelling of tissues in the throat and upper respiratory tract. High exposures may lead to pulmonary edema leading to hypoxemia and even death. Industrial exposure to nitrogen dioxide may cause genetic mutations, damage to developing foetus and decrease fertility in women. Industrial exposure to nitric oxide can cause unconsciousness, vomiting, mental confusion and damage to the teeth. So far, there is no evidence that nitrogen oxides are potential carcinogens.

Dietary Nitrate Supplementation and Exercise Performance

Nitrate, an inorganic anion abundant in vegetables, is converted *in vivo* to bioactive nitrogen oxides including NO. It is reported that nitrate most surprisingly decreases whole body oxygen during exercise with preserved or even enhanced maximal performance (Larsen et al., 2011; Jones, 2014), It is further reported that skeletal muscle mitochondria harvested after nitrate supplementation displayed an improvement in oxidative phosphorylation efficiency (P/O ratio) and a decrease in state of respiration with and without atractyloside and respiration without adenylates. The improved mitochondrial P/O ratio correlated to the reduction in oxygen cost during exercise. Nitrate reduced the expression of ATP/ADP ranslocase, a protein involved in proton conductance. It reported that dietary nitrate has profound effects on basal mitochondrial function also. These findings may have implications for exercise physiology- and lifestyle-related disorders that involve dysfunctional mitochondria (Larsen et al., 2011).

Treatment and Prevention

Treatment

Supplementations of ascorbic acid, methionine, alpha-tocopherol, methylene blue have been found to be effective in overcoming the problem of nitrate toxicity. These substances are effective in inhibiting nitrosation if present in gastric juices at all time in significant concentration.

Role of Antioxidants

As ROS are derived from a number of sources including nitrates, it was deemed feasible to readjust the balance of ROS production and detoxification by supplementing antioxidants. How to prevent or treat this oxidative stress? (Armitage et al., 2009). However, clinical trials testing this hypothesis showed little benefits in reducing respiratory, GIT, cardiovascular etc. events or mortality; some of the treatments even caused harm (Dotan et al., 2009).

Indeed, in vivo detoxification of increased ROS by antioxidant supplementation may be impossible due to the poor bioavailability of the antioxidants at the correct subcellular localization and at the optimal time point. Antioxidants may also promote new radical chain reactions initiated by their oxidized forms.

Importantly, the global removal of ROS may result in unwanted effects, since ROS also regulate important physiological functions. Therefore, the current strategy is to identify and target the major sources of ROS, including NADPH oxidases (NO_X). A major constraint in NADPH oxidase research and translation is the lack of specific inhibitors. All commonly used compounds, including apocynin and diphenyleneiodonium, are unspecific (Heumuller et al., 2008) and there are no specific NOX inhibitors commercially available. The novel NADPH oxidase inhibitor, VAS 2870, inhibits PDGF-stimulated cell migration and NADPH activity in vascular smooth muscle cells (Ten Freyhaus et al., 2006) as well as ox-LDL-stimulated ROS formation in endothelial cells (Stielow et al., 2006). However, its mechanism of action and selectivity for different NOX isoforms is unknown.

In vivo experiments with a small peptide inhibitor, gp91ds-tat, which interferes with the interaction between NOX2 and the regulator p47phox, had promising results, e.g. it reduced angioplasty-induced superoxide and neointimal hyperplasia of rat carotid artery (Jacobson et al., 2003). As individual NOX isoforms are regulated by different subunits, targeting these protein–protein interactions is one strategy for the development of isoform-specific NOX inhibitors.

For more details on NADPH oxidase inhibitors and their limitations, the published reviews (Cave, 2009; Williams and Griendling 2007; Lambeth et al., 2008; Selemidis et al., 2008) can be looked. Once specific inhibitors become available, their potential to manage CVD by preventing ROS formation can be fully explored. This concept may also be applicable to other chronic diseases associated with oxidative stress.

Hence, the dissection of the oxidative stress hypothesis into a detailed identification of clinically relevant enzymes targeted by ROS offers for the first time a more mechanistic approach to treatment of oxidative stress including CVD; eNOS recouplers/enhancers as well as sGC stimulators/activators target specific oxidative damage, which currently can only be detected at a functional level. Nevertheless, clinical efficacy and proof-of-concept animal models promise a far more positive outcome than for previous antioxidant approaches. Thinking forward, the NADPH oxidase field represents an uncharted territory in the management of CVD. We believe that the development of NO_X isoform selective inhibitors in the next decade will reveal the clinical relevance of this enzyme family. Twenty years after its discovery (Palmer et al., 1987), NO has come of age. It may be not so much NO synthesis itself but the signalling and oxidative events surrounding this unusual molecule that may produce the therapeutic outcomes that have been postulated for this highly protective vascular pathway and for management of oxidative stress including CVD.

Prevention

At environmental level:

Avoid Water Pollution

- Minimize the contamination of water supplies by nitrates originating from agricultural practice.
- Avoid inadvertent and excessive use of nitrogenous fertilizers.
- Avoid water pollution.
- Avoid the habit of open air defecation.
- Avoid the stagnation of waste water around the source of water.
- Avoid the sewage disposal directly to groundwater table.

Removal of Nitrate from Drinking Water

Denitrification: The removal of nitrate from water is costly and difficult to implement and maintain both at domestic as well as community level. The following are the known processes of denitrification:

- Physiochemical Process: The commonly used processes under this category are (i) Reverse osmosis (ii) Ion exchange (iii) Electrodialysis.
- Biological Treatments: This process is important since it bears similarity with metabolism of nitrates in living organisms.

At human level:

- Breast-feeding only, at least up to the age of 4 months.
- If top feed is necessary then preferably using of cow's milk (of course unadulterated) or dry milk preparations reconstituted with water containing low level of nitrate.
- Avoiding the use of high nitrate containing food for weaning.
- Avoiding use of nitrate and nitrite as preservative to the minimum extent possible especially to cured and canned meats and fish.
- Using of drinking water with low level of nitrates, especially for pregnant mothers.
- Avoiding WHO ORS preparation with locally available high nitrate water during diarrhoea.
- Avoiding long-term use of antiacid secretary agents especially in pregnant mothers and children. In case if it is necessary to use these drugs, they should be used cautiously and preferably with the use of antioxidants.

Reducing Nitrate Levels in Vegetables

The following measures will reduce nitrates in vegetables, however, some of them will reduce essential nutrients such as Vitamin C as well:

- Washing and cooking in water (nitrates are soluble in water).
- Peeling, e.g. in potatoes because the nitrates are concentrated in the skin and just below (again, this is part of failsafe eating).
- Discarding: in leafy vegetables such as lettuce, discarding the stem and mid-rib can decrease up to 40 per cent of the nitrates.
- A study of French fries showed that peeling reduced levels by 30 per cent; preheating and cutting reduced the nitrate content by a further 20 per cent and blanching by 30 per cent. After final frying only about five per cent of the original nitrate content remained (Reheating cold cooked vegetables has been shown not to increase nitrate levels as previously thought).
- For homegrown vegetables, choose low nitrate varieties (iceberg lettuce is one of the lowest and is failsafe); use slower releasing nitrogen sources such as manure and compost; avoid fertilizing just before harvest; grow vegetables outside if possible (lower light due to plastic or glass covering contributes to higher nitrates); harvest in full sun.
- A number of studies have suggested a protective effect of Vitamin C and other nutrients against either nitrates or particular cancers. Some experts think that the naturally occurring nutrients in vegetables are enough to protect against effects of nitrates. The WCRF warns: 'don't use high-dose supplements to protect against cancer'. Research shows that high-dose nutrient supplements can affect our risk of cancer, so it's best to opt for a balanced diet without supplements.

Conclusion

Nitrate and Nitrite are natural ions that are a part of the nitrogen cycle. Naturally occurring nitrate levels in surface and groundwater are generally a few milligrams per litre. In many groundwater, an increase of nitrate levels has been observed due to water percolating through nitrate rich rocks and owing to the farming practices of using chemical fertilizers.

For human consumption, WHO (2004) permits up to 50 mg/l but for short-term exposure only, whereas IS 10500 permits 45 mg/l (as NO_3) as desirable limit and 100 mg of NO_3 per litre as maximum permissible limit in the absence of alternate source.

Not much literature on human studies is available on nitrate toxicity except reports documenting methaemoglobinemia in infants due to high nitrate ingestion. Apart from methaemoglobinemia, few studies indicated nitrate as a cause of cancer, but it is still controversial and no firm conclusions have been drawn so far. The other effects observed were: increased infant mortality, abortions, birth defects, recurrent diarrhoea, recurrent stomatitis, early onset of hypertension, ischaemia, CVD, histopathological changes in cardiac muscles, alveoli of lungs and adrenal glands, recurrent respiratory tract infection in children, Nephotic syndrome in children, hypothyroidism and diabetes. Recent studies have indicated that high nitrate ingestion adversely affects the immune system of the body. Recently, an adaptation system to nitrate ingestion has also been reported. This adaptation to an enzyme cytochrome b_5 reductase has been shown to be protective to human being, but to a limited extent only.

More detailed epidemiological health related studies covering a large sample are required to provide insight into the nitrate toxicity on human beings. It is possible that a detailed study on pathophysiology of nitrate metabolism and its effect on human beings would yield a better understanding of the various diseases caused by nitrates and, their prevention and treatment.

Since it is very difficult and costly to remove nitrates from water because it is chemically non-reactive in dilute aqueous solutions, it is recommended that a simple change in habits and adoption of simple preventive measures may be urgently introduced in government campaign at least in high nitrate belts. Indiscriminate use of nitrogenous fertilizers should be avoided. The most important strategy is to promote breast-feeding up to the age of at least 6 months.

References

Allison, M.J. and Reddy, C.A. (1984). Adaptation of Gastrointestinal Bacteria in Response to the Changes in Dietary Oxalate and Nitrate. In: Klug M.J. and Reddy C.A. (eds) *Current Perspectives in Microbial Ecology, American Society for Microbiology*. Washington D.C. 248–256.

Anjana, Umar S., Iqbal, M. and Abrol, Y.P. (2007a) Are Nitrate Concentrations in Leafy Vegetables within Safe Limits? *Curr. Sci.* 92(3), 355–359.

Anjana, Umar S. and Iqbal, M. (2007b) Nitrate Accumulation in plants, Factors Affecting the Process, and Human Health Implications. A Review. *Agron. Sust. Develop.* 27, 45–57.

Anthea, M., Roshan, L., Hopkins, J., McLaughlin, C.W., Johnson, S., Warner, M.Q., LaHart, D. and Wright, J.D. (1993). *Human Biology and Health*. Englewood Cliffs, New Jersey, USA: Prentice Hall.

Arbuckle, T.E., Sherman, G.J., Corey, P.N., Walters, D. and Lo, B. (1988). Water Nitrates and CNS Birth Defects: A Population Based Case-Control Study. *Arch Environ Health*. 43: 162–167.

Armijo, R. and Coulson, A.H. (1975). Epidemiology of Stomach Cancer in Chile: The Role of Nitrogen Fertilizers. *Int J Epidemiol*. 4: 301–309.

Armijo, R., Orellana, M., Medina, E., Coulson, A.H., Sayre, J.W. and Detels, R. (1981). Epidemiology of Stomach Cancer in Chile-I: Case Control Study. *Int J Epidemiol*. 10: 53–56.

Armitage, M.E., Wingler, K., Schmidt, H.H. and La, M. (2009). Translating the Oxidative Stress Hypothesis into the Clinic: NOX versus NOS. *J Mol Med*. 87: 1071–1076.

Arnér, E.S. and Holmgren, A. (2000). Physiological Functions of Thioredoxin and Thioredoxin Reductase. *Eur J. Biochem*. 267 (20): 6102–6109.

Bartholomew, B. and Hill, M.J. (1984). The Pharmacology of Dietary Nitrate and the Origin of Urinary Nitrate. *Food Chem Toxicol*. 22: 789–795.

Benz, E.J. (2001). Hemoglobenopathies, Hemoglobin's with Altered Oxygen Affinity. In: Braunwald E., Fauci A.S., Kasper, D.L., Hauser, S.L., Longo, D.L. and Jameson, J.L. (eds) *Harrison's Principles of Internal Medicine 1*, 15th Ed. McGraw-Hill, New York. 1524.

Berger, H.M., Moison, R.M.W. and Van Zoeren-Grobben, D. (1997). Pathophysiology of Respiratory Distress Syndrome. In: David, T.J., (ed), *Recent Advances in Pediatrics*, New Delhi, Churchill livingstone, New Delhi. 14: 117–119

Bernstein, J.A., Alexis, N., Barnes, C., Bernstein, I.L., Bernstein, J.A., Nel, A., Peden, D., Diaz-Sanchez, D., Tarlo, S.M. and Williams, P.B. (2004). Health Effects of Air Pollution. *J Allergy Clin Immunol*. 114: 1116–1123.

Bodansky, O. (1951). Methemoglobinemia and Methemoglobin-producing Compounds. Phamacol Rev. 3: 144–196.

Burney, S., Caulfield, J.L., Niles, J.C., Wishnok, J.S. and Tannenbaum, S.R. (1999). The Chemistry of DNA Damage from Nitric Oxide and Peroxynitrite. *Mutat Res*. 424: 37–49.

Carmichael, P. and Lieben, J. (1963). Sudden Death in Explosive Workers. *Arch Environ Health*. 7: 424–439.

Cave, A. (2009). Selective Targeting of NADPH Oxidase for Cardiovascular Protection. *Curr Opin Pharmacol*. 9: 208–213.

CDC (Centers for Disease Control and Prevention) (1996). Spontaneous Abortions Possibly Related to Ingestion of Nitrate-contaminated Well Water-LaGrange County, Indiana, 1991–1994. *Morb Mortal Wkly Rep*. 45: 569–572.

Chaudhary, V., Kumar, M., Sharma, M. and Yadav, B.S. (2010). Fluoride, Boron and Nitrate Toxicity in Ground Water of North-West Rajasthan, India. *Environ Monit Assess.* 161: 343–348.

Cole, J.A. and Brown, C.M. (1980). Nitrate Reduction to Ammonia by Fermentative Bacteria: A Short Circuit in the Biological Nitrogen Cycle. *FEMS Microbiol Let.* 7: 65–72.

Colvin, C., Tredoux, G., Clarke, S., Le Maitre, D., Englebrecht, P. and Maherry, A. (2008). Integration Goes Underground: A Review of Groundwater Research in Support of Sustainable Development in South Africa. *Science Real and Relevant:* 2nd CSIR International Convention Centre, Pretoria, 17–18th November 2008. 10.

Comly, H.H. (1945). Cyanosis in Infants Caused by Nitrates in Well Water. *JAMA.* 129: 112–116.

Committee on Nitrate Accumulation, Ag. Board, Div. of Biology and Agriculture, National Research Council (1972). Hazards of Nitrate, Nitrite, and Nitrosoamines to Man and Livestock. In: Accumulation of Nitrate, National Academy of Sciences, 46–75, Washington DC.

Corbett, J.A., Wang, J.L., Hughes, J.H., Wolf, B.A., Sweetland, M.A., Lancaster, J.R. Jr, McDaniel, M.L. (1992a). Nitric Oxide and Cyclic GMP Formation Induced by Interleukin 1 Beta in Islets of Langerhans. Evidence for an Effective Role of Nitric Oxide in Islet Dysfunction. *Biochem J.* 287 (1), 229–235.

Corbett, J.A., Wang, J.L., Sweetland, M.A., Lancaster, J.R. Jr, McDaniel, M.L. (1992b). Interleukin 1 Beta Induces the Formation of Nitric Oxide by Beta-cells Purified from Rodent Islets of Langerhans. Evidence for the Beta-cell as a Source and Site of Action of Nitric Oxide. *J Clin Invest.* 90: 2384–2391.

Cornblath, M. and Hartmann, A.F. (1948). Methemoglobinemia in Young Infants. *J Pediat.* 33: 421–425.

Corre, W.J. and Breimer, T. (1979). Nitrate and Nitrite in Vegetables. Literature Survey nr. 39, Center for Agricultural Publishing Documentation, Wageningen.

Croen, L.A., Todoroff, K. and Shaw, G.M. (2001). Maternal Exposure to Nitrate from Drinking Water and Diet and Risk for Neural Tube Defects. *Am J Epidemiol.* 153: 325–331.

Cross, A.J., Pollock, J.R. and Bingham, S.A. (2003). Heam, not Protein or Inorganic Iron, is Responsible for Endogenous Intestinal N-nitrosation arising from Red Meat. *Cancer Res.* 63: 2358–2360.

Cuello, C., Correa, P., Haenszel, W., Cordillo, G., Brown, C., Archer, M. and Tannenbaum, S. (1976). Gastric Cancer in Colombia I. Cancer Risk and Suspect Environmental Agents. *J Natl Cancer Inst.* 57: 1015–1020.

Dahlquist, G.G., Blom, L.G., Persson, L.A., Sandstrom, A.I. and Wall, S.G. (1990). Dietary Factors and the Risk of Developing Insulin Dependent Diabetes in Childhood. *BMJ.* 300: 1302–1306.

Dallinga, J.W., Pachen, D.M., Lousberg, A.H., Van Geel, J.A., Houben, G.M., Stockbrugger, R.W., Van Maanen, J.M. and Kleinjans, J.C. (1998). Volatile N-nitrosamines in Gastric Juice of Patients with Various Conditions of the Gastrointestinal Tract Determined by Gas Chromatography-mass Spectrometry and Related to Intragastric pH and Nitrate and Nitrite Levels. *Cancer Lett.* 124: 119–125.

De Kok, T.M., Engels, L.G., Moonen, E.J. and Kleinjans, J.C. (2005). Inflammatory Bowel Disease Stimulates Formation of Carcinogenic N-nitroso Compounds. *Gut.* 54: 731.

De Roos, A.J., Ward, M.H., Lynch, C.F. and Cantor, K.P. (2003). Nitrate in Public Water Supplies and the Risk of Colon and Rectum Cancers. *Epidemiology.* 14: 640–649.

Deanfield, J., Donald, A., Ferri, C., Giannattasio, C., Halcox, J., Halligan, S., Lerman, A., Mancia, G., Oliver, J.J., Pessina, A.C., Rizzoni, D., Rossi, G.P., Salvetti, A., Schiffrin, E.L., Taddei, S. and Webb, D.J. (2005). Endothelial Function and Dysfunction. Part I: Methodological Issues for Assessment in the Different Vascular Beds: A Statement by the Working Group on Endothelin and Endothelial Factors of the European Society of Hypertension. *J Hypertens.* 23: 7–17.

Dolby, J.M., Webster, A.D., Borriello, S.P., Barclay, F.E., Bartholomew, B.A. and Hill, M.J.(1984). Bacterial colonization and nitrite concentration in the achlorhydric stomachs of patients with primary hypogammaglobulinaemia or classical pernicious anaemia. *Scand J Gastroenterol.* 19: 105–110.

Dorsch, M.M., Scragg, R.K., McMichael, A.J., Baghurst, P.A. and Dyer, K.F. (1984). Congenital Malformations and Maternal Drinking Water Supply in Rural South Australia: A Case-Control Study. *Am J Epidemiol.* 119: 473–486.

Dotan, Y., Pinchuk, I., Lichtenberg, D. and Leshno, M. (2009). Decision Analysis Supports the Paradigm that Indiscriminate Supplementation of Vitamin E Does More Harm Than Good. *Arterioscler Thromb Vasc Biol.* 29: 1304–1309.

Druckrey, H., Ivankovic, S. and Preussmann, R. (1966). Teratogenic and Carcinogenic Effects in the Offspring After Single Injection of Ethylnitrosourea in Pregnant Rats. *Nature.* 210: 1378–1379.

Duguid, J.B. (1960). The Thrombogenic Hypothesis and Its Implications. Postgrad. *Med J.* 36: 226–229.

Eatough, D.J., Benner, C.L., Bayona, J.M., Richards, G., Lamb, J.D., Lee, M.L., Lewis, E.A. and Hansen, L.D. (1989). Chemical Composition of Environmental Tobacco Smoke. Gas-Phase Acids and Bases. *Environ Sci Technol.* 23: 679–687.

ECETOC. (1988). Nitrate and Drinking Water. European Chemical Industry Ecology and Toxicology Centre, Technical Report No. 27, Brussels, Belgium.

Eizirik, D.L., Delaney, C.A., Green, M.H., Cunningham, J.M., Thorpe, J.R., Pipeleers, D.G., Hellerstrom, C. and Green, I.C. (1996) Nitric Oxide Donors Decrease the Function and Survival of Human Pancreatic Islets. *Mol Cell Endocrinol.* 118: 71–83.

Enami, S., Hoffmann, M.R. and Colussi, A.J. (2009). Absorption of Inhaled NO(2). *J Phys Chem B.* 113: 7977–7981.

Eskiocak, S., Dundar, C., Basoglu, T. and Altaner, S. (2005). The Effects of Taking Chronic Nitrate by Drinking Water on Thyroid Functions and Morphology. *Clin Exp Med.* 5: 66–71.

Eskiocak ,S. (1995). The Effects of Administration Chronic Nitrate by Drinking Water on Thyroid Hormones, Thyroidal Radioiodine Uptake and Mass. Thesis in Biochemistry Specialty. Samsun, Turkey.

Fan, A.M., Willhite, C.C. and Book, S.A. (1987). Evaluation of the Nitrate Drinking Water Standard with Reference to Infant Methaemoglobinemia and Potential Reproductive Toxicity. *Regul Toxicol Pharmacol.* 7: 135–148.

Farinati F., Della Libera, G., Cardin, R., Molari, A., Plebani, M., Rugge, M., Di Mario, F. and Naccarato, R. (1996). Gastric Antioxidant, Nitrites, and Mucosal Lipoperoxidation in Chronic Gastritis and Helicobacter Pylori Infection. *J Clin Gastroenterol.* 22: 275–281

Farinati, F., Lima, V., Naccarato, R. and Garro, A. (1989). J. Mutagenic Activity in Gastric Juice and Urine of Subjects with Chronic Atrophic Gastritis, Gastric Epithelial Dysplasia and Gastric Cancer. *Cancer Lett.* 48: 169–175.

Febo, A. and Perrino, C. (1995). Measurement of High Concentration of Nitrous acid Inside Automobiles. *Atmos Environ.* 29: 345–351.

Fewtrell, L. (2004). Drinking-Water Nitrate, Methemoglobinemia, and Global Burden of Disease: A Discussion. *Environ Health Perspect.* 112: 1371–1374.

Finlayson-Pitts, B.J. and Pitts, J.N. (2000). *Chemistry of the Upper and Lower Atmosphere-Theory, Experiments, and Applications*, Academic Press, San Diego.

Finlayson-Pitts, B.J., Wingen, L.M., Sumner, A.L., Syomin, D. and Ramazan, K.A. (2003). The Heterogeneous Hydrolysis of NO2 in Laboratory Systems and in Outdoor and Indoor Atmospheres: An Integrated Mechanism. *Phys Chem Chem Phys.* 5: 223–242.

Food and Drug Administration (1972). Teratologic Evaluation of FDA 71–7 (Sodium Nitrate). Washington, DC: US Department of Health and Human Services, Public Health Service, Food and Drug Administration, Publication No. PB 221775.

Fraser, P., Chilvers, C., Beral, V. and Hill, M.J. (1980). Nitrate and Human Cancer: A Review of the Evidence. *Int J Epidemiol.* 9: 3–11.

Gangolli, S.D., Van den Brandt, P.A., Feron, V.J., Janzowsky, C., Koeman, J.H., Speijers, G.J.A., Speigelhalder, B., Walker, R. and Wishnok, J.S. (1994). Nitrate, Nitrite and N-nitroso Compounds. *Eur J Pharmacol.* 292: 1–38.

Garg, H.K. and Bryan, N.S. (2009). Dietary Sources of Nitrite as a Modulator of Ischemia/Reperfusion Injury. *Kidney Int.* 75: 1140–1144.

Gepts, W. (1965). Pathologic Anatomy of the Pancreas in Juvenile Diabetes Mellitus. *Diabetes.* 14: 619–633

Givelber, H.M. and DiPaolo, J.A. (1969). Teratogenic Effects of N-ethyl-N-nitrosourea in the Syrian Hamster. *Cancer Res.* 29: 1151–1155.

Goodman, A.L., Underwood, G.M. and Grassian, V.H. (1999). Heterogeneous Reaction of NO2: Characterization of Gas-phase and Adsorbed Products from the Reaction, 2NO2(g) + H2O(a) ? HONO(g) + HNO3(a) on Hydrated Silica Particles. *J Phys Chem A.* 103: 7217–7223.

Gray, R., Peto, R., Brantom, P. and Grasso, P. (1991). Chronic Nitrosamine Ingestion in 1040 Rodents: The Effect of the Choice of Nitrosamine, The Species Studied, and the Age of Starting Exposure. *Cancer Res.* 51: 6470–6491.

Graziewicz M., Wink, D.A. and Laval, F. (1996). Nitric Oxide Inhibits DNA Ligase Activity: Potential Mechanisms for NO-mediated DNA Damage. *Carcinogenesis.* 17: 2501–2505.

Green, L.C., Ruiz de Luzuriaga, K., Wagner, D.A., Rand, W., Istfan, N., Young, V.R. and Tannenbaum, S.R. (1981). Nitrate Biosynthesis in Man. *Proc Natl Acad Sci USA.* 78: 7764–7768.

Gruener, N. and Shuval, H.I. (1970). Health Aspects of Nitrates in Drinking Water. In: Developments in Water Quality Research, Ann Arbor Humphrey Science Publishers, *Ann Arbor*, London. 89–106.

Gupta, A.B. and Kumar, S. (1999). Aerobic Denitrification: Myth or Reality? *Indian Journal of Environmental Sciences.* 3: 85–90.

Gupta, S.K., Fitzgerald, J.F., Chong, S.K.F., Croffie, J.M. and Garcia, J.G. (1998). Expression of Inducible Nitric Oxide Synthase (iNOS) mRNA in Inflamed Esophageal and Colonic Mucosa in Pediatric Population. *Am J Gastroenterol.* 93: 795–798.

Gupta, S.K., Gupta, R.C., Gupta, A.B., Seth, A.K., Bassin, J.K., Gupta, A. and Sharma, M.L. (2001). Recurrent Diarrhea in Areas with High Nitrate in Drinking Water. *Arch Environ Health.* 56: 369–373.

Gupta, S.K., Gupta, R.C., Gupta, A.B., Seth, A.K., Bassin, J.K., Gupta, A. (2000). Methaemoglobinaemia in Areas with High Nitrate Concentration in Drinking Water. *Natl Med J India.* 13: 58–61.

Gupta, S.K., Gupta, R.C., Seth, A.K., Gupta, A.B., Bassin, J.K., Gupta, D.K. and Sharma, S. (1999a). Epidemiological Evaluation of Recurrent Stomatitis, Nitrates in Drinking Water and Cytochrome b5 Reductase Activity. *Am J Gastroenterol.* 94: 1808–1812.

Gupta, S.K., Gupta, R.C., Gupta, A.B., Seth, A.K., Bassin, J.K., Gupta, A. (1999b). Adaptation of Cytochrome-b5 Reductase Activity and Methaemo-Globinaemia in Areas with a High Nitrate Concentration in Drinking-water. *Bull World Health Organ.* 77: 749–753.

Gupta, S.K., Gupta, R.C., Seth, A.K., Gupta, A.B., Sharma, M.L. and Gupta, A. (1999c). Toxicological Effects of Nitrate Ingestion on Cardiorespiratory Tissues in Rabbit. *S Asian J Prev Cardio.* 2: 101–105.

Hegesh, E. and Shiloah, J. (1982). Blood Nitrates and Infantile Methemo-Globinemia. *Clin Chim Acta.* 125: 107–115.

Helgason, T. and Jonasson, M.R. (1981). Evidence for a Food Additive as a Cause of Ketosis-prone Diabetes. *Lancet.* 2: 716–720.

Heumuller, S., Wind, S., Barbosa-Sicard, E., Schmidt, H.H., Busse, R., Schroder, K. and Brandes, R.P. (2008). Apocynin is not an Inhibitor of Vascular NADPH Oxidases But an Antioxidant. *Hypertension.* 51: 211–217.

Hill, M.J., Hawksworth, G. and Tattersall, G. (1973). Bacteria, Nitrosamines and Cancer of the Stomach. *Br J Cancer.* 28: 562–557.

Hirokawa, J., Kato, T. and Mafuné, F. (2008). Uptake of Gas-phase Nitrous Acid by pH-controlled Aqueous Solution Studied by a Wetted Wall Flow Tube. *J Phys Chem A.* 112: 12143–12150.

Höring, H. (1992). The Influence of Environmental Chemicals on the Thyroid. Bundesgesundheitsblatt. 35: 194–197.

Höring, H., Nagel, M. and Haerting, J. (1991). The Nitrate-dependent Endemic Thyroid Areas. In *Quantitative Methoden in der Epidemiologie.* Guugenmoos-Holzmann I (ed.), Springer, Berlin. 147–153.

Hsia, C.C. (1998). Respiratory Function of Hemoglobin. *N Engl J Med.* 338: 239–248.

Hughes, J.H., Colca, J.R., Easom, R.A., Turk, J. and McDaniel, M.L. (1990). Interleukin 1 Inhibits Insulin Secretion from Isolated Rat Pancreatic Islets by a Process that Requires Gene Transcription and mRNA translation. *J Clin Invest.* 86: 856–863.

Hughes, K.J., Meares, G.P., Chambers, K.T. and Corbett, J.A. (2009). Repair of Nitric Oxide-damaged DNA in â-cells requires JNK-dependent GADD45â Expression. *J Biol Chem.* 284: 27402–27408.

IARC. (1978). IARC Monographs on the Evaluation of Carcinogenic Risks of Chemicals to Man. Volume 17: Some N-Nitroso Compounds. World Health Organization.

International Agency for Research on Cancer. (2004). Tobacco Smoke and Involuntary Smoking, IARC Monographs (International Agency for Research on Cancer, Lyon, (France). 83.

IS 10500 (1995). Drinking Water Specification. Bureau of Indian standards, New Delhi, pp. 3.

Jacobson, G.M., Dourron, H.M., Liu, J., Carretero, O.A., Reddy, D.J., Andrzejewski, T. and Pagano, P.J. (2003). Novel NAD(P)H Oxidase Inhibitor Suppresses Angioplasty-induced Superoxide and Neointimal Hyperplasia of Rat Carotid Artery. *Circ Res.* 92: 637–643.

Jaffé, E.R. (1981). Methemoglobinemia. *Clin Haematol.* 10: 99–122.

Jahreis, G., Schone, F., Lüdke, H. and Hesse, V. (1987). Growth Impairment Caused by Dietary Nitrate Intake Regulated via Hypothyroidism and Decreased Somatomedin. *Endocrinol. Exp.* 21: 171–180.

Jaiswal, M., LaRusso, N.F., Nishioka, N., Nakabeppu, Y. and Gores, G.J. (2001). Human Ogg1, A Protein Involved in the Repair of 8-oxoguanine, is Inhibited by Nitric Oxide. *Cancer Res.* 61: 6388–6393.

Jaskiewicz, K., Van Helden, P.D., Wiid, I.J., Steenkamp, H.J. and Van Wyk, M.J. (1990). Chronic Atrophic Gastritis, Gastric pH, Nitrites and Micronutrient Levels in a Population at Risk for Gastric Carcinoma. *Anticancer Res.* 10: 833–836.

Jones, A.M. (2014). Dietary Nitrate Supplementation and Exercise Performance. *Sports Med 44* (Suppl 1).

Katti, S.R. and Sathyanesan, A.G. (1987). Lead Nitrate Induced Changes in the Thyroid Physiology of the Catfish Clarias Batrachus (L). *Ecotoxicol Environ Saf.* 13: 1–6.

Knotek, Z. and Schmidt, P. (1964). Pathogenesis, Incidence and Possibilities of Preventing Alimentary Nitrate Methemoglobinemia in Infants. *Pediatrics.* 34: 78–83.

Knox, E.G. (1972). Anencephalus and Dietary Intakes. *Br J Prev Soc Med.* 26: 219–223.

Kostraba, J.N., Gay, E.C., Rewers, M. and Hamman, R.F. (1992). Nitrate Levels in Community Drinking Waters and Risk of IDDM. *An Ecological Analysis. Diabetes* Care. 15: 1505–1508.

Kozliuk, A.S., Kushnir, G.V., Anisimova, L.A., Shroit, I.G. and Opopol, N.I. (1989). Immunological Status of Children Living in a Region with an Increased Level of Nitrates in the Drinking Water. *Gig Sanit.* 54: 19–22.

Kumar, S., Gupta, A.B., Gupta, S. (2002). Need for Revision of Nitrates Standards for Drinking Water: A Case Study of Rajasthan. *Indian J Environ Health.* 44: 168–72.

Kuper, F. and Til, H.P. (1995). Subchronic Toxicity Experiments with Potassium Nitrite in Rats. In: Health Aspects of Nitrate and its Metabolites (particularly nitrite). Proceedings of an International Workshop, Bilthoven (Netherlands), 8–10 November 1994. *Council of Europe Press.* 195–212.

Lambeth, J.D., Krause, K.H. and Clark, R.A. (2008). NOX Enzymes as Novel Targets for Drug Development. *Semin Immunopathol.* 30: 339–363.

Larsen, F.J., Schiffer, T.A., Borniquel, S., Sahlin, K., Ekblom, B.M., Lundberg, J.O. and Weitzberg, E. (2011). Dietary Inorganic Nitrate Improves Mitochondrial Efficiency in Humans. *Cell Metabolism.* 13(2): 149–159.

Leaf, C.D., Wishnok, J.S. and Tannenbaum, S.R. (1989). L-arginine is a Precursor for Nitrate Biosynthesis in Humans. *Biochem Biophys Res Commun.* 163: 1032–1037.

Lee, K., Xue, J., Geyh, A.S., Ozkaynak, H., Leaderer, B.P., Weschler, C.J. and Spengler, J.D. (2002). Nitrous Acid, Nitrogen Dioxide and Ozone Concentrations in Residential Environments. *Environ Health Persp.* 110: 145–149.

Longnecker, M.P. and Daniels, J.L. (2001). Environmental Contaminants as Etiologic Factors for Diabetes. *Environ Health Perspect.* 109: 871–876.

Malberg, J.W., Savage, E.P. and Osteryoung, J. (1978). Nitrates in Drinking Water and the Early Onset of Hypertension. *Environ Pollut.* 15: 155–160.

Malik, RPS. (2000). Agriculture and Water Quality in India Towards Sustainable Management. In: Water Quality Management and Control of Water Pollution. 73–85.

Mandrup-Poulsen, T. (1996). The Role of Interleukin-1 in the Pathogenesis of IDDM. *Diabetologia.* 39: 1005–1029.

Marshall, R.O., Disburger, H.J., MacVicar, R. and Hallmark, G.D. (1953). Studies on the Effect of Aeration on Nitrate Reduction by Pseudomonas Species Using N15. *J Bacteriol.* 66: 254–258.

Marshall, W. and Marshall, C.R. (1945). Action of Nitrites on Blood. *J Biol Chem.* 158: 187–208.

Matt, G.E., Matt, G.E., Quintana, P.J., Hovell, M.F., Bernert, J.T., Song, S., Novianti, N., Juarez, T., Floro, J., Gehrman, C., Garcia, M. and Larson, S. (2004). Households Contaminated by Environmental Tobacco Smoke: Sources of Infant Exposures. *Tob Control.* 13: 29–37.

Mensinga, T.T., Speijers, G.J. and Meulenbelt, J. (2003). Health Implications of Exposure to Environmental Nitrogenous Compounds. *Toxicol Rev.* 22: 41–51.

Mertes, S. and Wahner, A. (1995). Uptake of Nitrogen Dioxide and Nitrous Acid on Aqueous Surface. *J Phys Chem.* 99: 14000–14006.

Michaud, D.S., Mysliwiec, P.A., Aldoori, W., Willett, W.C. and Giovannucci, E. (2004). Peptic Ulcer Disease and the Risk of Bladder Cancer in a Prospective Study of Male Health Professionals. *Cancer Epidemiol Biomarkers Prev.* 13: 250–254.

Mirvish, S.S. (1983). The Etiology of Gastric Cancer: Intragastric Nitrosamide Formation and Other Theories. *J Natl Cancer Inst.* 71: 631–647.

Mirvish, S.S. (1991). The Significance for Human Health of Nitrate, Nitrite, and N-nitroso Compounds. In Nitrate Contamination: Exposure, Consequence, and Control, Bogardi I, Kuzelka RD and Ennenga WG (eds). NATO ASI Series, Springer Verlag. Berlin. 30: 253–266.

Mirvish, S.S. (1995). Role of N-nitroso Compounds (NOC) and N-nitrosation in Etiology of Gastric, Esophageal, Nasopharyngeal and Bladder Cancer and Contribution to Cancer of Known Exposures to NOC. *Cancer Lett.* 93: 17–48.

Misler, S., Barnett, D.W., Gillis, K.D. and Pressel, D.M. (1992). Electrophysiology of Stimulus-secretion Coupling in Human Beta-cells. *Diabetes.* 41: 1221–1228.

Møller, H., Landt, J., Pedersen, E., Jensen, P., Autrup, H. and Jensen, O.M. (1989). Endogenous Nitrosation in Relation to Nitrate Exposure from Drinking Water and Diet in a Danish Rural Population. *Cancer Res.* 49: 3117–3121.

Mondal, N.C., Saxena, V.K. and Singh, V.S. (2008). Occurrence of Elevated Nitrate in Ground Waters of Krishna Delta, *India. Afr J Environ Sci Technol.* 2: 265–271.

Morton, W.E. (1971). Hypertension and Drinking Water Constituents in Colorado. *A Am J Public Health.* 61: 1371–1378.

Mueller, R.L., Hagel, H.J., Wild, H., Ruppin, H. and Domschke, W. (1986). Nitrate and Nitrite in Normal Gastric Juice. Precursors of the Endogenous N-nitroso Compound Synthesis. *Oncology.* 43: 50–53.

Muhrer, M.E., Garner, G.B. and Pfander, W.H. (1959). The Effect of Nitrate on Reproduction and Lactation. *J An Sci.* 15: 1291–1292.

Murray, K.F. and Christie, D.L. (1993). Dietary Protein Intolerance in Infants with Transient Methaemoglobinemia and Diarrhea. *J Pediatr.* 122: 90–92.

Mustacich, D. and Powis, G. (2000). Thioredoxin Reductase. *Biochem J.* 346: 1–8.

Nagel, R.L. (2007). Methemoglobinemias and unstable hemoglobins. In *Cecil Text Book of Medicine*, Goldman L, Ausiello DA, Arend W, Armitage JO, Clemmon D, Drazen J, Griggs R, LaRusso N, Newman J, Foster E (eds). 23th. 1226–1227.

NAS. (National Academy of Sciences) Safe Drinking Water Committee (1977). Drinking Water and Health. Washington, D.C.

NAS. (National Academy of Sciences) Environmental Studies Board (1978). Nitrates: An Environmental Assessment. Washington, DC.

NAS. (National Academy of Sciences) (1981). The health effects of nitrate, nitrite, and N-nitroso compounds. Part 1 of a two-part study by the Committee on Nitrite and Alternative Curing Agents in Food. Report by the US National Research Council, National Academy of Sciences. Washington, DC, National Academy Press.

Niles, J.C., Wishnok, J.S. and Tannenbaum, S.R. (2006). Peroxynitrite-Induced Oxidation and Nitration Products of Guanine and 8-oxoguanine: Structures and Mechanisms of Product Formation. *Nitric Oxide.* 14: 109–121.

Nishimoto, I.N., Hanaoka, T., Sugimura, H., Nagura, K., Ihara, M., Li, X.J., Arai, T., Hamada, G.S., Kowalski, L.P. and Tsugane, S. (2000). Cytochrome P450 2E1 Polymorphism in Gastric Cancer in Brazil: Case-Control Studies of Japanese Brazilians and Non-Japanese Brazilians. Cancer Epidemiol Biomarkers Prev. 9: 675–680.

NRC - National Research Council. (1995). Nitrate and Nitrite in Drinking Water. National Research Council. Subcommittee on Nitrate and Nitrite in Drinking Water. National Academy Press, Washington, DC.

Paliwal, K.V. (1989). Pollution of Surface and Ground Waters. In: Water Pollution and Management. Varshney CK (Ed). Wiley Eastern Limited, New Delhi. 36–60.

Palmer, R.M., Ferrige, A.G. and Moncada, S. (1987). Nitric Oxide Release Accounts for the Biological Activity of Endothelium-derived Relaxing factor. *Nature.* 327: 524–526.

Parslow, R.C., McKinney, P.A., Law, G.R, Staines, A., Williams, R. and Bodansky, H.J. (1997). Incidence of Childhood Diabetes Mellitus in Yorkshire, Northern England, is Associated with Nitrate in Drinking Water: An Ecological Analysis. *Diabetologia.* 40: 550–556.

PHED Habitat Survey. (1991–93). Nitrate Concentration in Water in Villages of Rajasthan. Rajasthan by Public health Engineering Department of Rajasthan.

Pitts, J.N. Jr, Biermann, H.W., Tuazon, E.C., Green, M., Long, W.D. and Winer, A.M. (1989). Time-resolved Identification and Measurement of Indoor air Pollutants by Spectroscopic Techniques: Gaseous Nitrous Acid, Methanol, Formaldehyde and Formic Acid. *JAPCA.* 39: 1344–1347.

Pitts, J.N. Jr, Grosjean, D., Van Cauwenberghe, K., Schmid, J.P. and Fitz, D.R. (1978). Photooxidation of Aliphatic Amines Under Simulated Atmospheric Conditions: Formation of Nitrosamines, Nitramines, Amides and Photochemical Oxidant. *Environ Sci Technol.* 12: 946–953.

Pomeranz, A., Korzets, Z., Vanunu, D., Krystal, H. and Wolach, B. (2000). Elevated Salt and Nitrate Levels in Drinking Water Cause an Increase of Blood Pressure in Schoolchildren. *Kidney Blood Press Res.* 23: 400–403.

Porter, W.P., Jaeger, J.W. and Carlson, I.H. (1999). Endocrine, Immune, and Behavioral Effects of Aldicarb (carbamate), Atrazine (triazine) and Nitrate (fertilizer) Mixtures at Groundwater Concentrations. *Toxicol Ind Health.* 15: 133–150.

Prakasa Rao, EVS. and Puttanna, K. (2006). Strategies for Combating Nitrate Pollution. *Curr Sci.* 91: 1335–1339.

Rabinovitch , A. and Suarez-Pinzon, W.L. (1998). Cytokines and Their Roles in Pancreatic Islet Beta-cell. *Biochem Pharmacol.* 55: 1139–1149.

Reed, P.I., Smith, P.L., Haines, K., House, F.R. and Walters, C.L. (1981). Gastric Juice N-nitrosamines in Health and Gastro-duodenal Disease. *Lancet.* 2: 550–552.

Reed, P.I., Summers, K., Smith, P.L.R.., Walters, C.L., Bartholomew, B.A., Hill, M.J. et al. (1983). Effect of Ascorbic Acid treatment on Gastric Juice Nitrite and N-nitroso Compound Concentration in Achlorhydric Subjects. 24: 1492–1493.

Revenga, C. and Mock, G. (2001). Dirty Water: Pollution Problems Persist. EarthTrends: The Environmental Information Portal, Source: Pilot Analysis of Global Ecosystems: Freshwater Systems.

RIVM. (1989). Integrated Criteria Document Nitrate, Effects. Appendix to RIVM Report No. 758473012. Bilthoven, Rijksintituut voor de Volksgenzondheid en Milieuhygiëne (National Institute of Public Health and Environmental Protection) (RIVM Report No. A758473012).

Roberts, O.L., Holmes, K., Müller, J., Cross, D.A. and Cross, M.J. (2009). ERK5 and the Regulation of Endothelial Cell Function. *Biochem Soc Trans.* 37: 1254–1259.

Roediger, W.E., Deakin, E.J., Radcliffe, B.C. and Nance, S. (1986). Anion Control of Sodium Absorption in the Colon. *QJ Exp Physiol.* 71: 195–204.

Royal Commission on Environmental Pollution (RCEP) 7th Report (1979). Agriculture and Pollution, London, Her Majesty's Stationery Office, 87–125, London.

Ruddell, W.S., Bone, E.S., Hill, M.J. and Walters, C.L. (1978). Pathogenesis of Gastric Cancer in Pernicious Anaemia. *Lancet.* 1: 521–523.

Scarim, A.L., Heitmeier, M.R. and Corbett, J.A. (1997). Irreversible Inhibition of Metabolic Function and Islet Destruction after a 36-hour Exposure to Interleukin-1beta. *Endocrinology.* 138: 5301–5307.

Scheidleder, A., Grath, J., Winkler, G., Stark, U., Koreimann, C., Gmeiner, C., Nixon, S., Casillas, J., Gravesen, P., Leonard, J. and Elvira, M. (1999). *Groundwater Quality and Quantity in Europe.* European Environment Agency, Copenhagen.

Schmitz, J.T. (1961). Methemoglobinemia: A Cause of Abortions?. Preliminary Report. *Obstet Gynecol.* 17: 413–415.

Schwiede, M., Duijnisveld, W.H.M. and Bottcher, J. (2005). Investigation of Processes Leading to Nitrate Enrichment in Soils in the Kalahari Region, Botswana. *Physics and Chemistry of the Earth*, Parts A/B/C. 30: 712–716.

Scientific Committee on Food (SCF). (1995). Opinion on Nitrate and Nitrite, Annex 4 to Document III/5611/95, European Commission (ed.), Brussels, 20.

Selemidis, S., Sobey, C.G., Wingler, K., Schmidt, H.H. and Drummond, G.R. (2008). NADPH Oxidases in the Vasculature: Molecular Features, Roles in Disease and Pharmacological Inhibition. *Pharmacol Ther.* 120: 254–291.

Shrivastava, V.K., Katti, S.R. and Sathyanesan, A.G. (1987). Effect of Lead Nitrate on Thyroid Function of the Indian Palm Squirrel, Funambulus Pennanti (Wroughton). *Bull Environ Contam Toxicol.* 38: 981–984.

Shuval, H.I. and Gruener, N. (1972). Epidemiological and Toxicological Aspects of Nitrates and Nitrites in the Environment. *Am J Public Health.* 62: 1045–1052.

Simon, C., Bostedt, H. and Adams, W. (2000). Juvenile Goiter in a Goat Herd in Northwest Germany. *Schweiz Arch Tierheilkd.* 142: 339–347.

Singer, B.C., Hodgson, A.T. and Nazaroff, W.W. (2003). Gas-phase Organics in Environmental Tobacco Smoke: 2. Exposure-relevant Emission Factors and Indirect Exposure from Habitual Smoking. *Atmos Environ.* 37: 5551–5561.

Singer, B.C., Revzan, K.L., Hotchi, T., Hodgson, A.T. and Brown, N.J. (2004). Sorption of Organic Gases in a Furnished Room. *Atmos Environ.* 38: 2483–2494.

Singh, B., Singh, Y. and Sekhon, G.S. (1994). Fertilizer N Use Efficiency and Nitrate Pollution of Groundwater in Developing Countries. 15th World Congress on Soil Science. 5a: 174–191.

Sinha, A.K., Seth, G. and Batheja, K. (2008). Nitrate and Fluoride Contamination in Groundwater Churu Block, Rajasthan. *Journal of Indian Water Works Association.* 40: 45–49.

Sleiman, M., Gundel, L.A., Pankow, J.F., Jacob, P., Singer, B.C. and Destaillats, H. (2010). Formation of Carcinogens Indoors by Surface-mediated Reactions of Nicotine with Nitrous Acid, Leading to Potential Thirdhand Smoke Hazards. *Proc Natl Acad Sci.* 107: 6576–6581.

Sloop, G.D. and Garber, D.W. (1997). The Effects of Low-density Lipoprotein and High-density Lipoprotein on Blood Viscosity Correlate with Their Association with Risk of Atherosclerosis in Humans. *Clin. Sci.* 92: 473–9.

Smith, A.D., Datta, S.P. and Smith, G.H. (1997). Nitric Oxide. In: Smith, A.D., Datta, S.P., Smith, G.H. et al., (eds) *Oxford Dictionary of Biochemistry and Molecular Biology*, Oxford University Press, Oxford, New York-Tokyo. 451.

Smith, R.L., Ceazan, M.L., Myron, N.D. and Brooks, H. (1994). Autotrophic, Hydrogen-Oxidizing, Denitrifying Bacteria in Groundwater, Potential Agents for Bioremediation of Nitrate Contamination. *Applied and Environmental Microbiology.* 60: 1949–1955.

Southern, C., Schulster, D. and Green, I.C. (1990). Inhibition of Insulin Secretion by Interleukin 1-beta and Tumor Necrosis Factor-alpha via an L-arginine-dependent Nitric Oxide Generating Mechanism. *FEBS Lett.* 276: 42–44.

Spalding, R.F. and Exner, M.E. (1993). Occurrence of Nitrate in Groundwater: A Review. *Journal of Environmental Quality.* 22(3): 392–402.

Speijers, G.J.A. (1996). Nitrate, in: World Health Organization (eds), Toxico Logical Evaluation of Certain Food Additives and Contaminants in Food. *Food Additive Series 35,* Geneva, 325–360.

Spengler, J.D., Brauer, M., Samet, J.M. and Lambert, W.E. (1993). Nitrous Acid in Albuquerque, New Mexico, Homes. *Environ Sci Technol.* 27: 841–845.

Steer, S.A., Scarim, A.L., Chambers, K.T. and Corbett, J.A. (2006). Interleukin-1 Stimulates Beta-cell Necrosis and Release of the Immunological Adjuvant HMGB1. *PLoS Med.* 3: e17.

Stielow, C., Catar, R.A., Muller, G., Wingler, K., Scheurer, P., Schmidt, H.H. and Morawietz, H. (2006). Novel Nox inhibitor of oxLDL-induced Reactive Oxygen Species Formation in Human Endothelial Cells. *Biochem Biophys Res Commun.* 344: 200–205.

Stokinger, H.E. (1982). Aliphatic Nitro Compounds, Nitrates, Nitrites. In: G.D. Clayton and F.E. Clayton (eds) *Patty's Industrial Hygiene and Toxicology,* Vol. 2A, John Wiley & Sons, New York. 4169–4201.

Sund, J.M., Wright, M.J. and Simon, J. (1957). Weeds Containing Nitrate Cause Abortion in Cattle. *Agronomy Journal.* 49: 278–279.

Super, M., Heese, H.V., MacKenzie, D., Dempster, W.S. and du Plessis, J.J. (1981). An Epidemiological Study of Well-water Nitrates in a Group of South West African/Namibian Infants. *Water Research.* 15: 1265–1270.

Szaleczky, E., Prónai, L., Nakazawa, H. and Tulassay, Z. (2000). Evidence of in Vivo Peroxynitrite Formation in Patients with Colorectal Carcinoma, Higher Plasma Nitrate/Nitrite Levels, and Lower Protection Against Oxygen Free Radicals. *J Clin Gastroenterol.* 30: 47–51.

Szokeova, E., Tajtakova, M., Mirossay, L., Mojzis, J., Langer, P., Marcinova, E., Petrovicova, J., Zemberova, E. and Bodnar, J. (2001). Effect of Nitrates on Active Transport of Iodine. *Vnitr Lek.* 47: 768–771.

Tajtakova, M., Semanova, Z., Tomkova, Z., Szokeova, E., Majoros, J., Radikova, Z., Sebokova, E., Klimes, I. and Langer, P. (2006). Increased Thyroid Volume and Frequency of Thyroid Disorders Signs in Schoolchildren from Nitrate Polluted Area. *Chemosphere.* 62: 559–564.

Tamir, S., Burney, S. and Tannenbaum, S.R. (1996). DNA Damage by Nitric Oxide. *Chem Res Toxicol.* 9: 821–827.

Tannenbaum, S.R. (1983). N-nitroso Compounds: A Perspective on Human Exposure. *Lancet 1:* 629–632.

Ten Freyhaus, H., Huntgeburth, M., Wingler, K., Schnitker, J., Baumer, A.T., Vantler, M., Bekhite, M.M., Wartenberg, M., Sauer, H. and Rosenkranz, S. (2006). Novel Nox Inhibitor VAS2870 Attenuates PDGF Dependent Smooth Muscle Cell Chemotaxis, But Not Proliferation. *Cardiovasc Res.* 71: 331–341.

Til, H.P., Falke, H.E., Kuper, C.F. and Willems, M.I. (1988). Evaluation of the Oral Toxicity of Potassium Nitrite in a 13-week Drinking-water Study in Rats. *Food Chem Toxicol.* 26: 851–859.

Toledo, J.C.Y., Tanus-Santos, J.E., Sabha, M., Cittadino, M., Buonalumi, H., Moreno, H. and Yugar-Toledo, J.C. (2004). Uncontrolled Hypertension, Uncompensated Type II Diabetes, and Smoking Have Different Patterns of Vascular Dysfunction. *Chest.* 125: 823–830.

USP DI (1990). *Nitrates Systemic.* 2033–2036.

Ustyugova, I.V., Zeman, C., Dhanwada, K. and Beltz, L.A. (2002). Nitrates/Nitrites Alter Human Lymphocyte Proliferation and Cytokine Production. *Arch Environ Contam Toxicol.* 43: 270–276.

Van Duijvenboden, W. and Matthijsen, A.J.C.M. (1989). Integrated Criteria Document Nitrate. Blithoven, Rijsinstiuut voor de Volksgezondheid en Milieuhygiene (National Institute of Public Health and Environmental Protection) (RIVM report No. 758473012).

Van Maanen, J.M., Van Dijk, A., Mulder, K., de Baets, M.H., Menheere, P.C., Van der Heide, D., Mertens, P.L. and Kleinjans, J.C. (1994). Consumption of Drinking Water With High Nitrate Levels Causes Hypertrophy of the Thyroid. *Toxicol Lett.* 72: 365–374.

Vander, A.J., Sherman, J.H., Luciano, D.S. (1994). The Digestion and Absorption of Food. *Human Physiology.* New York: McGraw-Hill, Inc, New York. 561–600.

Verge, C.F., Howard, N.J., Irwig, L., Simpson, J.M., Mackerras, D. and Silink, M. (1994). Environmental Factors in Childhood IDDM. A Population-based, Case-control Study. *Diabetes Care.* 17: 1381–1389.

Vermeer, I.T., Engels, L.G., Pachen, D.M., Dallinga, J.W., Kleinjans, J.C. and Van Maanen, J.M. (2001). Intragastric Volatile N-nitrosamines, Nitrite, pH, and Helicobacter Pylori During Long-term Treatment with Omeprazole. *Gastroenterology*. 121: 517–525.

Vermeer, I.T., Pachen, D.M., Dallinga, J.W., Kleinjans, J.C. and Van Maanen, J.M. (1998). Volatile N-nitrosamine Formation after Intake of Nitrate at the ADI Level in Combination with an Amine-rich Diet. *Environ Health Perspect*. 106: 459–463.

Vladeva, S, Gatseva, P. and Gopina, G. (2000). Comparative Analysis of Results from Studies of Goitre in Children from Bulgarian Villages with Nitrate Pollution of Drinking Water in 1995 and 1998. *Cent Eur J Public Health*. 8: 179–181.

Violante, A., Cianetti, A. and Ordine, A. (1973). Adrenal Cortex Function During Subacute Poisoning with Sodium Nitrite. *Quad Sclavo Diagn*. 9: 907–920.

Virtanen , S.M., Jaakkola, L., Rasanen, L., Ylonen, K., Aro, A., Lounamaa, R., Akerblom, H.K. and Tuomilehto, J. (1994). Nitrate and Nitrite Intake and the Risk for Type 1 Diabetes in Finnish Children. Childhood Diabetes in Finland Study Group. *Diabet Med*. 11: 656–662.

Wainman, T., Weschler, C.J., Lioy, P. and Zhang, J. (2001). Effects of Surface Type and Relative Humidity on the Production and Concentration of Nitrous Acid in a Model Indoor Environment. *Environ Sci Technol*. 35: 2200–2206.

Weisenberg, E., Grauer, F., Baron, Y. and Rajstein, J. (1982). An Assessment of the Implication of Nitrate-rich Water on Salivary Nitrate. *Ecotoxicol Environ Saf*. 6: 97–101.

Weisenburger, D.D. (1990). Environmental Epidimiology of Non-Hodgkin's Lymphoma in Eastern Nebraska. *Am J Ind Med*. 18: 303–305.

Weyer, P.J., Cerhan, J.R., Kross, B.C., Hallberg, G.R., Kantamneni, J., Breuer, G., Jones, M.P., Zheng, W. and Lynch, C.F. (2001). Municipal Drinking Water Nitrate Level and Cancer Risk in Older Women: The Lowa Women's Health Study. *Epidemiology*. 12: 327–338.

WHO. (1977a). Nitrate, Nitrite and N-Nitroso Compounds (1977). *Environmental Health Criteria 5*, Geneva. 15.

WHO. (1977b). Nitrate, Nitrite and N-Nitroso Compounds. *Environmental Health Criteria 5*. Geneva. 82.

WHO (1977c). Nitrate, Nitrite and N-Nitroso Compounds. *Environmental Health Criteria 5*. Geneva. 73.

WHO (1977d). Nitrate, Nitrite and N-Nitroso Compounds. *Environmental Health Criteria 5*, Geneva. 78.

WHO (1977e). Nitrate, Nitrite and N-Nitroso Compounds. *Environmental Health Criteria 5*, Geneva. 37.

WHO (1985a). Health hazards from Nitrate in Drinking-water. In report on a WHO Meeting, Copenhagen, 5–9 March 1984. Copenhagen, WHO Regional Office for Europe (Environmental Health Series No. 1). Copenhagen. 14–16.

WHO (1985b). Health Hazards from Nitrate in Drinking-water. In report on a WHO Meeting, Copenhagen, 5–9 March 1984. Copenhagen, WHO Regional Office for Europe (Environmental Health Series No. 1), Copenhagen. 23–40.

WHO (1985c). Health Hazards from Nitrate in Drinking-water. In report on a WHO Meeting, Copenhagen, 5–9 March 1984. Copenhagen, WHO Regional Office for Europe (Environmental Health Series No. 1), Copenhagen. 19.

WHO (1996). Toxicological Evaluation of Certain Food Additives and Contaminants. Prepared by the Forty-Fourth Meeting of the Joint FAO/WHO Expert Committee on Food Additives (JECFA). Geneva, World Health Organization, International Programme on Chemical Safety (WHO Food Additives Series 35).

WHO (2003). Nitrate and Nitrite in Drinking-water. Background Document for Development of WHO Guidelines for Drinking-water Quality. World Health Organization, Geneva.WHO/SDE/WSH/04-03/56. English.

WHO (2004). Chemical Aspects. In Guidelines for Drinking Water Quality. Geneva. 1: 145–196.

Williams, H.C. and Griendling, K.K. (2007). NADPH Oxidase Inhibitors: New Antihypertensive Agents?. *J Cardiovasc Pharmacol.* 50: 9–16.

Wilson, G.L., Mossman, B.T. and Craighead, J.E. (1983). Use of Pancreatic Beta Cells in Culture to Identify Diabetogenic N-nitroso Compounds. *In Vitro.* 19: 25–30.

Winterbourn, C.C., McGrath, B.M. and Carrell, R.W. (1976). Reaction Involving Superoxide and Normal and Unstable Haemoglobins. *Biochem J.* 155: 493–502.

Xu, G., Song, P. and Reed, P.I. (1992). The Relationship Between Gastric Mucosal Changes and Nitrate Intake via Drinking Water in a High Risk Population for Gastric Cancer in Moping County, China. *Eur J Cancer Prev.* 1: 437–443.

Yabushita, A., Enami, S., Sakamoto, Y., Kawasaki, M., Hoffmann, M.R. and Colussi, A.J. (2009) Anion-catalyzed Dissolution of NO_2 on Aqueous Microdroplets. *J Phys Chem A.* 113: 4844–4848.

Zaki, A., Ait Chaoui, A., Talibi, A., Derouiche, A.F., Aboussaouira, T., Zarrouck, K., Chait, A. and Himmi, T. (2004). Impact of Nitrate Intake in Drinking Water on the Thyroid Gland Activity in Male Rat. *Toxicol Lett.* 147: 27–33.

Zaldivar, R. (1977). Nitrate Fertilizers as Environmental Pollutants, Positive Correlation Between Nitrates ($NaN0_3$ and $KN0_3$) used Per Unit Area and Stomach Cancer Mortality Rates. *Experientia.* 33: 264–265

Zhang, H., Luo, Y., Zhang, W., He, Y., Dai, S., Zhang, R., Huang, Y., Bernatchez, P., Giordano, F.J., Shadel, G., Sessa, W.C. and Min, W. (2007) Endothelial-specific Expression of Mitochondrial Thioredoxin Improves Endothelial Cell Function and Reduces Atherosclerotic Lesions. *Am J Pathol.* 170: 1108–1120.

10

Socio-Economic and Health Impacts of Waterlogging and Water Pollution in IGNP, Stage I Canals

A.N. Arora and Rohit Goyal

Introduction of large scale surface water irrigation projects in arid and semi arid regions has brought positive changes like revolution in agricultural production by increasing it manyfold, uptrend in economy and urbanization along with negative ones which include waterlogging and salinization in the region. Because of the twin problems of waterlogging and salinization, full advantage of the costly irrigation projects is not being realized. Waterlogging and salinity affect farm productivity, which leads to decrease in household income and thereby, to lower standards of living. This in turn, affects the social and economic conditions of that part of the farming community, which is affected by waterlogging and salinity. This is also true for the irrigation water of Indira Gandhi Nahar Pariyojna (IGNP), which irrigates Hanumangarh and Sriganganagar along with other districts of Rajasthan. These districts started facing these problems soon after introduction of the canal system, and the problem further aggravated after diversion of Ghaggar river floodwaters to 18 depressions through Ghaggar Diversion Channel (GDC) in year 1967–68. The appearance of waterlogging conditions began from *Badopal* area due to topographic set up and extended all along the Drishadwati bed. There were 907.5

ha of waterlogged area (water table 0–1.0 m bgl) in year 1968 (Sharma and Dhiman, 1994), which increased to 8,600 ha in 90–91. The area continued to increase up to 1997–98 to the extent of 22,008 ha and then due to shortage of water in canals and drought years decreased to 5,755 ha in year 2002–03.

The impact of waterlogging in canal command area is reflected not only in the agro-sector but also in the socio-economic status of the living population in the area. A detailed socio-economic survey can provide very useful information in this regard. In the past, such socio-economic surveys were done, with the intention of assessing positive/beneficial aspects of irrigation projects; however, not much information is available on negative aspects of waterlogging and how it affects the socio-economic pattern.

Baseline socio-economic surveys if conducted at the planning stage can provide a better comparison base for assessing positive and negative socio-economic impacts of the project at different post-project stages with pre-project conditions. These surveys will also provide a better insight to planners and policy makers for minimizing the degradation in social status of farmers of the command area by adopting various water management strategies at the planning stage itself. Gathering data for such a plan provides the opportunity of integrating the necessary benchmark studies into the planning process, which is also a requirement of World Bank's Guidelines.

Literature Review

Large investments have been made the world over in expanding the area under irrigation, which in turn has made a significant contribution to world food production. In dry arid regions, rainfed agriculture produces very few crops, but irrigation facilitates a manyfold increase in the agricultural productivity of lands. Irrigation in the arid and dry regions very often leads to waterlogging and salinization. There is also an apprehension that salinization of land is inevitable as a result of which irrigation schemes can have only a finite life and cannot be sustained indefinitely.

Sharma and Dhiman (1994) conducted a study on canal irrigation and environmental degradation in Indira Gandhi Canal in parts of Sriganganagar district and concluded that introduction of large-scale irrigation in this arid region has brought in positive environmental changes viz. improvement of microclimate, changes in floral elements, microrelief variations, revolution in economy and urbanization along with the negative ones which include soil salinity, waterlogging and marshy lands. Safwat (1998) has stated while discussing water logging and soil salinity, that effect of rising water tables and salinity can be classified into three broad categories: decreased agricultural production, decline in ecosystem health, and infrastructure damage apart from other social losses. Scudder (1998) while discussing social impacts has pointed out that a major problem in irrigation projects is inadequacy of pre-project benchmark studies against which subsequent impacts can be assessed. In two such projects in Sri Lanka, such studies occurred after construction activities have began. The only solution to this situation is to initiate the necessary pre-project benchmark studies much earlier in the planning process. Shekhawat and Singh (2002), while conducting a socioeconomic survey in the area of Lakhuwali, an Indo-Dutch project (under network centre Hanumangarh), has stated that waterlogging and salinity affect farm productivity which leads to decrease in income, and thereby to lower standard of living of farmers. This in turn affects the social and economic conditions of the farming community. Baseline socioeconomic surveys in such pilot and other irrigation projects must be conducted along with other investigations at planning stage and such surveys shall be repeated after 3–4 years of implementation to know the socioeconomic impacts of the projects.

Case Study of Indira Gandhi Nahar Pariyojna Stage I

A study was conducted on socioeconomic impacts of waterlogging in Indira Gandhi Nahar Pariyojna Stage I in 14 waterlogged villages of Hanumangarh and Sriganganagar district of Rajasthan as shown in Figure 10.1. The study included demographic details, pre- and post-occupation, literacy profile, assets owned, income and expenditure profiles of households and other social impacts. The survey generated quantitative values of social and economic parameters related to the farming community of the waterlogged villages in the study area for pre- and post-waterlogged stages.

LISS III digital remote sensing data of IRS-IC satellite was used to delineate waterlogged and salinity affected area. A sample size of 7.5 per cent of affected families was considered as adequate in the

Figure 10.1

Study Area showing Waterlogged Area and Villages Surveyed

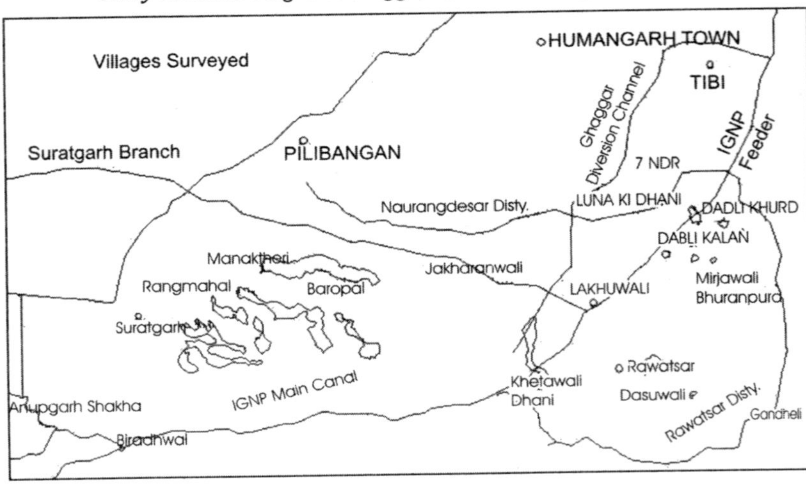

present study. A questionnaire was prepared in consultation with social scientists, economists, research organizations and government departments of Agriculture, Irrigation, Command Area Development, IGNP and Economics and Statistics. The face-to-face interview method was followed in the study to formalize input from the affected farmers. The questionnaire focuses on selected parameters viz. household composition, religion and category, literacy level, agricultural and residential landholdings, machinery and agricultural implements, live stock, cropping pattern, input use, properties damaged, income and expenditure profile of households affected by waterlogging and salinity.

Out of 280 households surveyed, 91.42 per cent were of Hindu religion, 1.43 per cent Muslim and 7.15 per cent of Sikh religion. Category-wise, 81.08 per cent belonged to OBC, 10.35 per cent General, 7.97 per cent to SC and 0.60 per cent to ST category. A marked change was noticed in female literacy rate (13.17% to 45.64%) in spite of all odds due to waterlogging. As regards residential landholdings, number of kutcha houses increased from 63 to 111 and percentage of pucca houses were decreased by about 33 per cent as shown in Figure 10.2. In case of machinery and agricultural implements, the number of ploughs, carts and chaff-cutters was found to have decreased due to reduction in cultivated area and increased number of tractor-trolley, oil engines and jeeps as alternative for earning livelihood. The total number of livestock as well as owners was found considerably reduced, due to waterlogging conditions in the area.

Figure 10.2

Type of Households Pre/Post Waterlogging

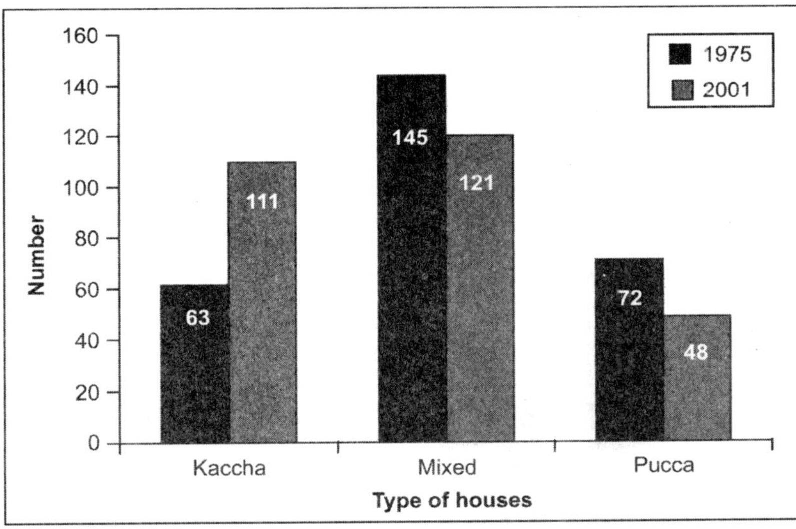

Average landholding of the households (280) was 6.60 ha. Out of total 280 households, 17.2 per cent were found in category of small farmers (landholding < 2 ha), 21.4 per cent in semi-medium category (landholding 2–4 ha), 11.4 per cent under large category (landholding > 10 ha) and remaining 50 per cent household fall under medium category (4–10 ha) Out of total land area of 1,867.14 ha, 1,286.5 ha was fully waterlogged and 72.15 per cent, 23.5 per cent and 4.28 per cent were having land in 1, 2 and 3 fragments respectively.

Details of submergence of properties as in year 2001 are presented in Table 10.1. Out of 280 residential lands, 189 were fully submerged, out of 1,855.25 ha of irrigated land 1,286.50 ha was also submerged, 29 well/tubewells were found submerged and out of 4,419 trees, 2,399 were fully submerged and 1,370 partially submerged.

Table 10.1

Details of Submerged/Affected Properties

Property	*Area/number*		*Condition NS/FS/PS 2001*		
			NS	*FS*	*PS*
Residential land	280	Nos.	91	189	-
Irrigated land	1,855.25	ha.	568.75	1286.50	-
Non-irrigated land	11.89	ha.	11.89	-	-
Wells/tubewells	177	Nos.	148	29	-
Trees	4,419	Nos.	650	2,399	1,370

Note: NS-Not submerged, FS-fully submerged, PS-Partially submerged.

A marked change was found in cropping pattern in pre- and post-waterlogging period in both Rabi and Kharif crops due to water logging of lands and market fluctuations in rates of food grains. Cropped areas of two major crops, wheat and cotton have decreased as shown in Figure 10.3.

Figure 10.3

Change in Cropping Pattern

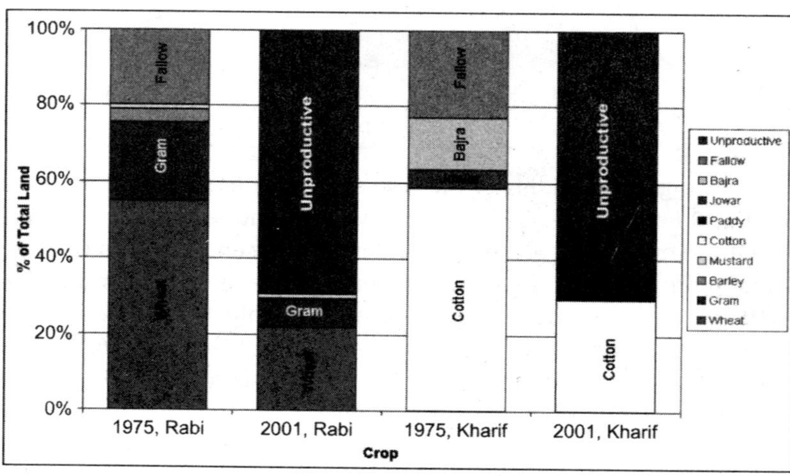

Average yield and gross return of crops per hectare was compared and it was found that net return from cotton was ₹ 5,274 in normal area as against ₹ 4,545 in partially waterlogged area. A major reduction in case of wheat was found from ₹ 6,985 in normal area to ₹ 2,521 in partially waterlogged area.

Questions were asked during household surveys about the annual household income and also a rough distribution of income from various sources such as agriculture, custom labour, livestock, artisan work etc. The average annual income per household and per capita income before and after waterlogging has been presented in Figure 10.4.

Figure 10.4

Annual Per Capita Income

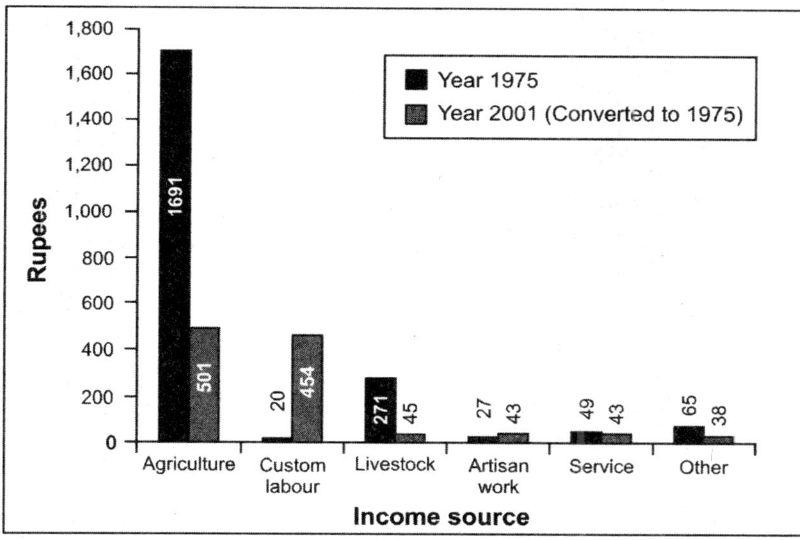

It can be seen that income under agriculture subhead has decreased considerably and income from custom labour, artisan work and services have increased. There is a considerable fall in per capita income, which has decreased by 47.16 per cent.

Statements given by the respondent farmers about their annual expenditure on household during household survey are compiled and analysed. Per capita expenditure under various subheads profile has been presented in Figure 10.5.

Annual household expenditure on food has decreased dramatically in terms of per capita expenditure. Annual household expenditures have gone up considerably in case of electricity, water, fuel and clothes. Similarly, per capita expenditure is nearly stable in case of education though it has increased in terms of household expenditure as well as in terms of percentage expenditure. There is a considerable increase in expenditure on health. Overall per capita

Figure 10.5

Per Capita Annual Expenditure under Various Subheads

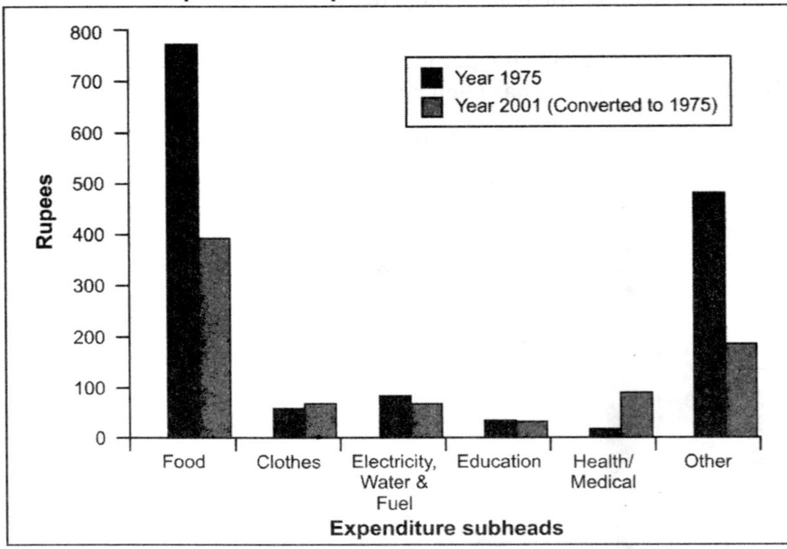

expenditure has also decreased due to decrease in overall income on account of waterlogging and salinity in the study area.

Social Impacts of Waterlogging and Salinity

Apart from the specific questions covered in the questionnaire, the farmers were asked about any other social impacts, not covered in the questionnaire, which they were experiencing. In response to this question, the statements given by the farmers have been compiled and presented in Table 10.2.

Table 10.2

Social Impacts of Waterlogging and Salinization

S. No.	*Statements*	*Number of Respondents*
1.	Our living standard has gone from bad to worst due to waterlogging	247
2.	Our social status has gone down considerably	232
3.	We are finding it very difficult to marry our sons and daughters. No family is willing to marry their sons and daughters in our waterlogged affected households	165

Cont'd...

...Cont'd

4.	We, the affected families are on the verge of breakdown. Continuous famine for last four years have worsened our condition	149
5.	Some of our family members are working as custom labour on daily wage basis	102
6.	We are doing some contract cultivation, but finding it difficult to meet even basic minimum needs of the family	92
7.	Unemployment and underemployment have completely upset our families. We are not getting alternative jobs in waterlogged villages	91
8.	Most of our family members are not maintaining good health due to malnutrition and costly medical treatment	87
9.	We are under heavy debt of moneylenders and banks due to non-payment of instalments	63
10.	There is a rise in family disputes and social conflicts due to heavy reduction in family earnings	45
11.	We are compelled to do some odd jobs like stitching clothes, selling milk, running a small shop, which do not match with our conventional job of agriculture	27
12.	Some of our family members though have switched over to artisan work like mason, carpenter, but finding it difficult to adjust	18
13.	We are earning our livelihood by running tractors and jeeps on hire after taking loans from banks	13
14.	If the present condition continues, some of us will be compelled to commit suicide like in Punjab and other states	6
15.	My wife ran away with other man due to financial stringencies.	1

Water Pollution

The Indira Gandhi Canal supplies drinking water to eight districts in north-west Rajasthan. It begins from Harike Barrage at the confluence of Sutlej and Beas rivers in Punjab and enters Hanumangarh. The Indira Gandhi Canal irrigates and supplies drinking water to Hanumagarh, Ganganagar, Bikaner, Churu, Nagaur, Jaisalmer, Barmer and Jodhpur. Zila Parishad, Hanumangarh had filed a petition in the National Green Tribunal (NGT) to pass an order directing the respondents, including the State of Punjab, and the Ministry of Environment and Forests, to take immediate steps to check pollution in Sutlej which is the main source of pollution of Indira Gandhi Canal because of a large number of industries situated in Ludhiana through which the river passes. The Parishad cited several national and international scientific studies which have drawn attention to the high levels of pollution in Sutlej and the Indira Gandhi Canal, and desired protection of fundamental, legal and customary rights of the people of Hanumangarh. It said that Punjab

Government had failed to perform the duties in accordance with law laid down by various statutes for checking water pollution.

The Indira Gandhi Canal is more polluted than the permissible limits of particulates, chemicals and biological organisms as shown by various studies conducted by the government bodies, government functionaries and private organizations. There are 1,332 hazardous waste generating industries in and around Ludhiana. The main industries are dyeing, electroplating and induction furnace manufacturing, generating 5,498 MT of waste. Water pollution can cause cancer, hepatitis, gastro-enteritis, jaundice and other circulatory, respiratory and skin diseases, the petition claims. There has been no decline in the number of follow-up cases either. In 2009, the Bikaner-based cancer hospital received 6,138 new cases and 45,357 follow-up cases. In 2010, there were 6,295 new cases and 45,189 follow-ups. In 2011, new cases decreased to 5,787 and follow-up cases to 43,189. However, 60 to 70 per cent of the new cases are from Punjab, the hospital authorities had confirmed in a report.

Conclusion

Baseline socio-economic surveys of major irrigation projects must be conducted at planning stage, so that a better comparison of positive and negative impacts of the project can be made at different post-project stages with pre-project conditions.

In mega irrigation projects, social impact assessment of beneficiaries due to displacement, rehabilitation, waterlogging and salinity, if perceived at planning stage, can provide a broader platform to decision makers in eliminating/minimizing the miseries of affected farmers due to negative impact of such projects.

A socio-economic survey of waterlogged villages in IGNP Stage I has revealed that social losses due to waterlogging and salinization in major irrigation projects are much more than the economic losses of land, agriculture and properties. Affected farmers apart from unemployment and underemployment are finding it difficult to manage day-to-day needs of the household. There is increase in family disputes and social conflicts and they have to accept some non-traditional odd jobs.

Canal water is used for drinking purpose by human beings and animals in these two districts. With water pollution by addition of untreated industrial and domestic waste by industries, local bodies of Ludhiana city in Punjab, people are suffering from severe diseases like cancer, etc. There is an immediate need of checking this pollution by

Punjab Government by installing sewage treatment plants (STPs) to add only treated water into river Satluj at Harike reservoir. Government of India and Central Pollution Control Board (CPCB) should intervene to give relief to population of affected districts.

Regular water quality monitoring at entry level in Rajasthan and at some points downstream also is a must, so that water quality deterioration may be monitored properly, to take steps for maintaining the water quality within permissible levels.

References

Arora, A.N., Goyal, R., Kashyap, R.P. (2005). Socio-economic Impacts of Irrigation Projects, Seminar Proceedings on 'Internal and External River Linking in Rajasthan'. 63–71.

Arora, A.N., Dhaneria, K.S., Goyal, Rohit and Kashyap, R.P. (2005). Socio-economic Impacts of Waterlogging: A Case Study of Indira Gandhi Nahar Pariyojna, Stage- I. *Journal of Indian Water Resource Society*. 25(1).

Natrajan, P.M. (2001). Wasted to the Sea, *The Hindu*, March 13.

Safwat, Abdel-Dayem (1998). Waterlogging and Soil Salinity, in *Water Resources: Environmental Planning, Management and Development*, A.K. Biswas (ed.), Tata McGraw Hill. 99–116.

Scudder, Thayer (1998). 'Social Impacts', *Water Resources: Environmental Planning, Management and Development,* Asit K. Viswas, Tata McGraw Hill Publishing Company. 623–665.

Sharma, H.S. and Dhiman, S.C. (1994). Canal Irrigation and Environmental Degradation: A Case Study of Indira Gandhi Canal in Parts of Ganganagar District. *Journal of Water and Landuse Management 3 (III and IV), DST India*. 41–55.

Shekhawat, R.S. and Singh, V.P. (2002). Socio-economic Base Line Survey of Indo-Dutch Project on Subsoil Drainage Research in IGNP Command at Lakhuwali, Rajasthan Agriculture University. 1–32.

Singh, Bharat (2004). Role of Water Resources Development in Economic Growth of India. *Journal of Indian Water Resources Society*. 24(2): 1–18.

Zila Parishad Haumangarh Petition (2014). National Green Tribunal about drinking water pollution in IGNP Canal.

11

Impact of Heavy Metals on Human Health

Assessment of Sanganer Area Textile Waste Water

Jaishree

Environmental pollution has become a worldwide phenomenon. Pollution of water bodies is a matter of concern in the developing countries including India. Industrial effluents, urban runoff, direct disposal of wastes into the water bodies, agricultural fertilizer and animal wastes remain the major water contaminants. It is also reported that textile and dyeing industry pose a major environmental threat because of the large amount of water and dyes involved in the manufacturing process. Large amount of chemically different dyes are employed for various industrial applications including textile dyeing. Textile industry is considered to be one of water polluters. Wastewater largely is a by-product and effluent from small, medium and large scale industries belonging to tie and dye, paper and pulp and blue pottery. Numbers of studies have been carried out both in the country and in other parts of the world (Afrin et al., 2014; Jaishree and Khan, 2014; Joshi and Satnani, 2012 and Marwari and Khan, 2012). In general, wastewater contains substantial amounts of beneficial nutrients and toxic heavy metals, which are creating opportunities and problems for agricultural production, respectively. Excessive accumulation of heavy metals in agricultural soils through wastewater irrigation, may not only result in soil contamination, but also lead to elevated heavy metal uptake by crops, and thus affect food quality and safety (Muchuweti et al., 2006).

Heavy metals enter into food from natural sources like soil, air and water through wastewater irrigation, solid waste disposal, mining, smelting, sludge applications, vehicular exhaust, fertilizers, fungicides and industrial activities (Muhammad et al., 2008; Radwan and Salama, 2006). Consumption of food crops contaminated with heavy metals is a major food chain route for human exposure (Khan et al., 2008). The absorption of heavy metals in the system varies to certain extent depending on various factors. Heavy metals, in general, are non-biodegradable, have long biological half-lives and have the potential for accumulation in the different body organs, leading to acute as well as chronic toxic effects (Radwan and Salama, 2006). The problem of heavy metal contamination is getting serious in different countries of the world, especially in developing countries. Moreover, as heavy metal bio-accumulation increases in nutrition deprived state, developing countries with higher prevalence of undernutrition are at a greater risk of heavy metal toxicity. The contamination of heavy metals in agricultural soils can pose health problems; therefore, there is a need to investigate the possible risks in the population due to the chronic exposure of heavy metal contamination in vegetables and fruits. Vegetables are common diet taken by population throughout the world, being sources of essential nutrients, antioxidants and metabolites (Thomsan and Kelly, 1990). Both essential and toxic elements are present in vegetables over a wide range of concentrations as they are said to be good absorber of metals from the soil (Lokeshwari and Chandrappa, 2006; Eslami et al., 2007). Vegetables absorb these metals from contaminated soils as well as from polluted environmental deposits through the roots and assimilate them into the edible part of plant tissues or deposit them on the surface of vegetables (Haiyan and Stuanes, 2003; Nawaeji, 2009).

Study Area

In the present study Sanganer, Jaipur has been selected as study area where textile wastewater drains into Amanishah Nalla. It is located between 26°49'N to 26°51'N latitude and 75°46'E to 75°51'E longitude (Figure 11.1). It is famous for tie and dye printing, waste paper recycling and blue pottery industries. These industries discharge untreated waste water into Amanishah Nalla (erstwhile Dravyawati River) in Sanganer (Figure 11.2). A large number of small, medium and large scale textile industrial units are located in Sanganer. The untreated waste water which contains chemicals like anelin, caustic soda, acids, bleaching powder and heavy metals is lifted through pumps and utilized in irrigating agricultural fields for growing vegetables and other crop plants (Figure 11.3).

Figure 11.1

Showing Study Area, Sanganer, Jaipur, Rajasthan, India

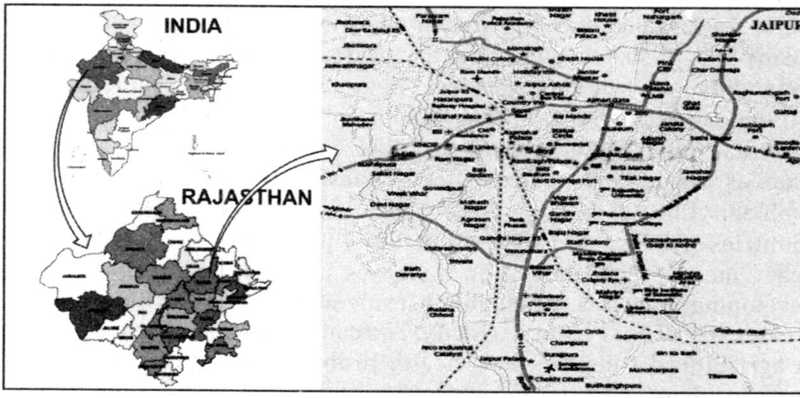

Figure 11.2

Showing Amanishah Nalla

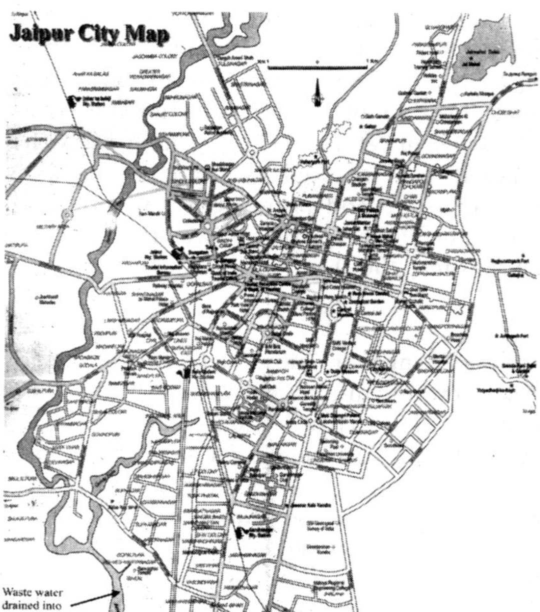

Figure 11.3

Series of Pump Sets Used to Pump Waste Water for Irrigating Agricultural Fields shown in the Background

Figure 11.4

Shows Effect of Waste Water on Hands and Feet of Workers Using Chemicals Without Gloves

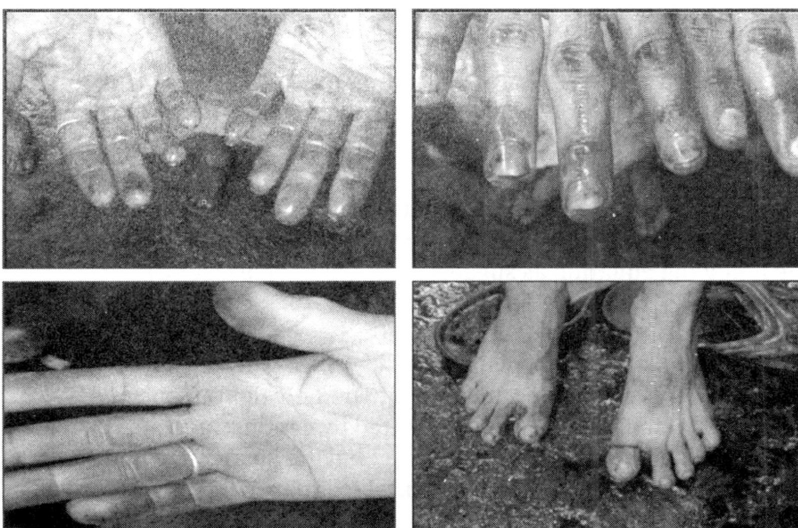

Material and Methods

Soil and vegetable samples were collected from five different fields of the study area at post-flowering stage. Analysis of samples was done for measuring concentration of heavy metals.

Heavy Metal Analysis by Atomic Absorption Spectrophotometer

Atomic Absorption Spectrophotometer (AAS Model GBC 932) was used for analysis of heavy metals in water namely Cobalt (Co), Zinc (Zn), Cadmium (Cd), Nickle (Ni), Copper (Cu), Lead (Pb) and Chromium (Cr). The 25 ml of the sample was digested in di-acid mixture of HNO_3 and per chloric acid in the ratio of (10:1). The digestion was performed in 100 ml conical flasks and to facilitate complete digestion, the samples in di-acid mixture were kept overnight at room temperature. These flasks containing samples and di-acid mixture were heated on hot plate until a clear solution was obtained. This was followed by a slow but complete evaporation of acids. Then, the volume of the digested samples was made up to 100 ml with the help of the double distilled water. Finally, these solutions were analysed by Atomic Absorption Spectrophotometer.

Data Analysis

Transfer Factor (TF)

Soil to plant metal transfer was computed as transfer factor(TF) using the equation

$$TF=C_{PLANT}/C_{SOIL}$$

Where, C Plant is the concentration of heavy metals in plants and C Soil is the concentration of heavy metals in soil.

Daily Intake of Metals (DIM)

Daily intake of vegetables in adult was calculated by data obtained during the survey through a questionnaire. DIM was calculated using the following equation as suggested by Chary et al. (2008).

$$DIM=C_{metal} \times C_{factor} \times D_{food\ intake}/B_{average\ weight}$$

Where, C_{metal}, C_{factor}, $D_{food\ intake}$, $B_{average\ weight}$ represent the heavy metal concentrations in plants (mg kg^{-1}), conversion factor (0.085), daily intake of vegetables and average body weight, respectively. The average vegetable intake was calculated by conducting a survey where about 200 people (males and females) having an average body weight of 48 kg was asked for their daily intake of particular vegetables from sampling sites.

Health Risk Index (HRI): To assess the human health risk of heavy metals, it is necessary to calculate the level of human exposure to that metal by tracing the route of exposure of pollutant to human body. There exist many exposure routes for heavy metals that depend upon a contaminated media of soil and vegetables on the recipients. Receptor population use the vegetables enriched with higher concentration of heavy metals which enter the human body, leading to health risks (Khan et al., 2008). In the present research work, vegetables grown wastewater were collected from the study area and their metal concentration was used to calculate the health risk index (HRI). The health risk index of the present research work was compared with the one reported by Khan et al. (2008) and Jan et al. (2010); results of HRI were found to be lower than both the earlier studies. Value of HRI depends upon the daily intake of metals (DIM) and oral reference dose (R_{fd}). R_{fd} is an estimated per day exposure of metal to the human body that has no hazardous effect during life time (USEPA, 2006). The health risk index for Cr, Ni, Cu, Pb, Cd, Mn and Zn by consumption of contaminated vegetables was calculated using following equation:

$$HRI = DIM/R_{fd}$$

Where DIM represents the daily intake of metals and R_{fd} represents reference oral dose. R_{fd} for Cr, Ni, Cu, Pb, Cd, Co and Zn is 1.5, 0.02, 0.04, 0.004, 0.001, 0.033 and0.30 (mg/kg bw/day) respectively.

Result and Discussion

Concentration of heavy metals in soil obtained from different agricultural sites of Sanganer was found to be significantly high. The concentration of heavy metals in soil was above the safe limits set by USEPA, WHO and Indian standards. The concentrations of heavy metals (mg/g) in agricultural soil collected from different crop production sites are presented in Table 11.1. Zn concentration varied from 11.758 mg/gm to 16.495 mg/gm, Cu concentration was from 11.013 mg/gm to 11.589 mg/gm, Ni was from 4.623 mg/gm to 5.011 mg/gm, Cd varied from 4.586 mg/gm to 5.234 mg/gm, Cr concentration varied from 6.189 mg/gm to 6.998 mg/gm, Pb was from 5.286 mg/gm to 5.896 mg/gm and Co varied from 3.684 mg/gm to 3.912 mg/gm.

Table 11.1

Concentration of Heavy Metals in Soil

Sites	*Heavy Metals (mg/gm)*						
	Zn	*Cu*	*Ni*	*Cd*	*Cr*	*Pb*	*Co*
Site 1	13.289	11.125	5.011	4.586	6.378	5.896	3.684
Site 2	15.236	11.589	4.789	4.865	6.189	5.286	3.712
Site 3	16.495	11.445	4.623	4.912	6.245	5.895	3.776
Site 4	11.758	11.512	5.121	5.019	6.539	5.319	3.845
Site 5	14.698	11.013	4.948	5.234	6.998	5.489	3.912

Concentration of heavy metals in cow pea on a dry weight basis grown in wastewater is presented in Table 11.2. Results reveal that the heavy metal concentration was significantly higher in cow pea. The heavy metals were estimated in mg/gm dry weight. In plants, accumulated Zn concentration varied from 15.684 mg/gm to 15.708 mg/gm, Cu concentration varied from 8.348 m/gm to 8.878 mg/gm, Ni varied from 6.098 mg/gm to 6.186 mg/gm, Cd concentration varied from 5.264 mg/gm to 5.446 mg/gm, Cr was varied from 7.475 mg/gm to 7.505 mg/gm, Pb concentration varied from 6.567 mg/gm to 6.595 mg/gm and Co concentration varied from 5.538 mg/gm to 5.858 mg/gm.

Table 11.2

Concentration of Heavy Metals in Vigna unguiculata (cow pea)

SITES	*Heavy Metals (mg/gm)*						
	Zn	*Cu*	*Ni*	*Cd*	*Cr*	*Pb*	*Co*
Site 1	15.708	8.742	6.109	5.264	7.475	6.575	5.538
Site 2	15.690	8.348	6.123	5.289	7.487	6.579	5.546
Site 3	15.684	8.568	6.098	5.312	7.490	6.567	5.548
Site 4	15.696	8.796	6.142	5.348	7.501	6.583	5.553
Site 5	15.689	8.878	6.186	5.446	7.505	6.595	5.858

Transfer Factor (TF), Daily Intake of Metal (DIM) and Health Risk Index (HRI)

Transfer factor for cow pea in different heavy metals: Zn was 1.097, Cu was 0.764, Ni was 1.251, Cd was 1.082, Cr was 1.157, Co was 1.481 and Pb was 1.179. Plants uptake the heavy metals from soil and then it accumulates in different parts of the plant (Table 11.3).

Values of DIM calculated for adults (average age 48 years), are presented in Table 11.3. It reveals that the values of daily intake of metal were high for vegetables grown at different five agricultural sites. The trend for DIM in vegetables grown was in the order of Zn^{2+}> Cu^{2+}>Cr^{3+}>Pb^{2+}>Ni^{2+}>Cd^{2+}>Co^{2+}. The highest value of DIM for Zn was 0.061 mg/kg/day and lowest value of DIM for Co was 0.008 mg/kg/day.

To assess the health risk associated with heavy metal contamination of plants grown locally, estimated exposure and risk index were calculated. The results showed that Cd, Pb and Ni contamination in plants had greatest potential to pose health risk to the consumers. HRI was maximum for Cd^{2+} in cow pea and minimum or out of danger for Cr^{2+}.

Table 11.3

Transfer Factor, Daily Intake of Metal and Health Risk Index

Heavy Metals	*Average Concentration (mg/gm)*	*Transfer Factor*	*HRI*	*DIM (mg/kg/day)*
Zn	15.689	1.097	0.203	0.061
Cu	8.666	0.764	0.85	0.034
Ni	6.131	1.251	0.5	0.010
Cd	5.331	1.082	9	0.009
Cr	7.491	1.157	0.019	0.029
Pb	6.579	1.179	3	0.012
Co	5.608	1.481	0.242	0.008

Heavy metals causes different types of diseases; Zn causes stomach cramps, nausea, vomiting, anaemia, damage to the pancreas, Ni causes stomach cramps, nausea, vomiting, anaemia, damage to the pancreas, Pb causes haematological and cardiovascular effects (hypertension), kidney damage, Cd causes damage to kidneys and lungs, fragile bones, anaemia, increased risk of cancer if inhaled, Co affects the respiratory sensitization, asthma, shortness of breath and decreased pulmonary function, Cr causes skin rashes, upset stomachs and ulcers, respiratory problems, weakened immune systems, kidney and liver damage, alteration of genetic material, lung cancer and Cu causes the hematological manifestations, such as myelodysplasia, ananemia, leukopenia (low white blood cell count) and neutropenia (low count of neutrophils, a type of white blood cell that is often called 'the first line of defense' for the immune system). Vegetables take up metals by absorbing them from contaminated soils, as well as from deposits on different parts of the vegetables exposed to the air from polluted environments (Sobukola et al., 2010). Prolonged human consumption of unsafe concentrations of heavy metals in food-stuffs may lead to the disruption of many biological and biochemical processes in the human body (WHO, 1993; Jarup, 2003). Intake of vegetables is an important path of heavy metal toxicity to human beings. Dietary intake of heavy metals through contaminated vegetables may lead to various chronic diseases. A number of serious health problems can develop as a result of excessive uptake of dietary heavy metals (Arora et al., 2008). Furthermore, the consumption of heavy metal-contaminated food can seriously deplete some essential nutrients in the body causing a decrease in immunological defenses, intrauterine growth retardation, impaired psycho-social behaviours, disabilities associated with malnutrition and a high prevalence of upper gastrointestinal cancer.

Conclusion

It is concluded that continuous wastewater irrigation in the agricultural land has caused an ample build up of toxic metals. Food items analysis and monitoring should be performed in order to prevent excessive build up of these heavy metals in the food chain. Variations in the heavy metal concentrations between the test vegetables/cereal crops reflect the differences in uptake capabilities and their further translocation to the edible portion of the plants. Heavy metals have a toxic impact, but detrimental impact becomes apparent only when long-term consumption is done. Consumption of foodstuff with elevated levels of heavy metals may lead to high level of accumulation in the body causing related health disorders. It is suggested that wastewater should be used only after adequate treatment.

References

Afrin, R., Mia, M.Y. and Akter, S. (2014). Investigation of Heavy Metals (Pb, Cd, Cr, Cu, Hg and Fe) of the Turag River in Bangladesh. *Journal of Environment Science and Natural Resources*. 7(2): 133–136.

Arora, M., Bala, K., Rani, S., Rani, A., Kaur, B. and Mittal, N. (2008). Heavy Metal Accumulation in Vegatables Irrigated with Different Water Sources. *Food Chem.* 111(4): 811–815.

Awashthi, S.K. (2000). Prevention of Food Adulteration Act No. 37 of 1954. Centre and State Rules as Amended for 1999, 3rd ed., Ashoka Law House, New Delhi.

Chary, N.S., Kamala, C.T. and Raj, D.S.S. (2008). Assessing Risk of Heavy Metals from Consuming Food Grown on Sewage Irrigated Soils and Food Chain Transfer. *Ecotoxicology Environment Safety*. 69: 513–524.

Eslami, A., Khaniki, Gh. R.J., Nurani, M., Meharasbi, M., Peyda, M. and Azimi, R. (2007). Heavy Metals in Edible Green Vegetables Grown along the Sites of the Zanjan Roads Iran. *Journal of Biological Sciences*. 7: 943–948.

European Union (2002). Heavy Metals in Wastes, European Commission on Environment. <http://ec.europa.eu/environment/waste/studies /pdf/ heavymetalsreport.pdf>.

Haiyan, W. and Stuanes, A. (2003). Heavy Metal Pollution in Air-water-soil-plant System of Zhuzhou City, Hunan Province, China. *Water, Air and Soil Pollution*. 147: 79–107.

Jaishree and Khan, T.I. (2014). Monitoring of Heavy Metal in Textile Waste Water of Sanganer, Jaipur (Rajasthan). *International Journal of Scientific and Research Publications.* 4(3): 1–4.

Jan, F.A., Ishaq, M., Khan, S., Ihsanullah, I., Ahmad, I. and Shakirullah, M. (2010). A Comparative Study of Human Health Risks via Consumption of Food Crops Grown on Wastewater Irrigated Soil (Peshawar) and Relatively Clean Water Irrigated Soil (lower Dir). *J. Hazard. Mater.* 179: 612–621.

Jarup, L. (2003). Hazards of Heavy Metal Contamination, *Br. Med. Bull.* 68: 167–182.

Joshi, V.J. and Santani D.D. (2012). Physicochemical Characterization and Heavy Metal Concentration in Effluent of Textile Industry. *Universal Journal of Environmental Research and Technology.* 2(2): 93–96.

Khan S., Cao Q., Zheng Y., Huang Y. and Zhu Y. (2008). Health Risks of Heavy Metals in Contaminated Soils and Food Crops Irrigated with Wastewater in Beijing, China. *Journal of Science of Food and Agriculture* (73): 446–454.

Khan, S., Cao, Q., Zheng, Y.M., Huang, Y.Z. and Zhu, Y.G. (2008). Health Risks of Heavy Metals in Contaminated Soils and Food Crops Irrigated with Wastewater in Beijing, China. *Environmental Pollution.* 152: 686–692.

Lokeshwari, H. and Chandrappa, G.T. (2006). Impact of Heavy Metal Contamination of Bellandur Lake on Soil and Cultivated Vegetation. *Current Science.* 91: 622–627.

Marwari, R. and Khan, T.I. (2012). Effect of Textile Wastewater on Tomato Plant, *Lycopersicon Esculentum. Journal of Environment and Biology.* 33: 849–854.

Muchuweti, M., Birkett, J.W., Chinyanga, E., Zvamya, R., Scrimshaw, M.D. and Lister, J.N. (2006). Heavy Metal Content of Vegetables Irrigated with Mixtures of Waste Water and Sewage Sludge in Zimbabwe: Implication for Human Health. *Agri. Ecosys. Environ.* 112: 41–48.

Muhammad, F., Farooq, A. and Umer, R. (2008). Appraisal of Heavy Metal Contents in Different Vegetables Grown in the Vicinity of an Industrial Area. *Pakistan Journal of Botany.* 40(5): 2099–2106.

Nwajei, G.E. (2009). Trace Elements in Soils and Vegetations in the Vicinity of Shell Petroleum Development Company Operating Area in Ughelli, Delta State of Nigeria. *Journal of Sustainable Agriculture.* 3: 574–578.

Radwan, A. and Salama, A. (2006). Market Basket Survey for Some Heavy Metals in Egyptian Fruits and Vegetables, *Food and Chemical Toxicology.* 44: 1273–1278.

Sobukola, O.P, Adeniran, O.M., Odedairo, A.A., Kajihausa, O.E. (2010). Heavy Metal Levels of Some Fruits and Leafy from Selected Markets in Lagos, Nigeria Vegetables, *African Journal of Food Science.* 4(2): 389–393.

Thompson, H.C. and Kelly, W.C. (1990). *Vegetable Crops.* 5th Edn. McGraw Hill Publishing Company, New Delhi.

US-EPA, IRIS (2006). United States, Environmental Protection Agency, Integrated Risk Information System. http://www.epa.gov/ iris/subst.

WHO (1993). Evaluation of Certain Food Additives and Contaminants. 41st Report of the Joint FAO/WHO Expert Committee on Food Additives. Technical Report Series. World Health Organisation, Geneva.

Section–IV
Sanitation and Health

12

Water Sanitation and Health Outcomes

Alpana Kateja and Sanjukta Saha

The global importance of water, sanitation and hygiene for development, poverty reduction and health is reflected in the United Nation's Millennium Declaration, in particular its seventh Millennium Development Goal (MDG). It talks about ensuring sustainability by reducing the proportion of people without sustainable access to safe drinking water and basic sanitation by half, by the year 2015 (United Nations, 2000). Poor sanitation, hygiene and water are responsible for about 50 per cent of the consequences of childhood and maternal underweight, primarily through the synergy between diarrhoeal diseases and undernutrition, whereby exposure to one increases vulnerability to the other (Hutton and Haller, 2004; World Bank, 2008; Blossner and Onis, 2005). WHO (2008) estimates that globally eighty-eight per cent of diarrhoea is attributable to unsafe water, inadequate sanitation or insufficient hygiene. It also states that almost one-tenth of global disease burden could be prevented by improving water supply, sanitation, hygiene and management of water resources. Another estimate reports that water, sanitation and hygiene are responsible for four per cent of all deaths and 5.7 per cent of total disability-adjusted life years (Pruss et al., 2008). In a 7-point action plan to reduce the childhood diarrhoeal morbidity and mortality, UNICEF and WHO (2009) have included at least three directly related to water and sanitation; (1) improved water supply quantity and quality, including treatment and safe storage of household water; (2)

promotion of hand washing with soap; and (3) community-wide sanitation promotion.

Regarding drinking water, sanitation and hygiene, the report of UNICEF and WHO (2012) Joint Monitoring Programme for Water Supply and Sanitation estimates the following facts for India:

(1) A considerable proportion of global population who have gained access to drinking water since 1990 live in India (522 million).

(2) India is still home to 97 million people without access to improved water supply. This represents about 12.58 per cent of global population that remains unserved.

(3) Since 1990, 251 million people have gained access to improved sanitation in India.

(4) Eleven countries constitute more than three-quarters (76%) of the global population without improved sanitation facilities. One-third of the 2.5 billion people without improved sanitation live in India (814 million). It warns that India is off track to meet the MDG sanitation target.

(5) Nearly 60 per cent of those practising open defecation live in India (626 million people).

In this backdrop our study tries to analyse overall situation of drinking water and sanitation in Rajasthan, one of India's Empowered Action Group states having miserable health scenario. The present paper also attempts to analyse whether the differences in the sanitary conditions could explain differential health outcomes. The paper, therefore, builds on the hypothesis that the districts where a significant proportion of population have access to safe drinking water and sanitation have achieved better health outcomes. The analysis is carried out for 2011, census data which has very recently come out in houses, household amenities and assets tables of Census of India, 2011. For data on mortality rates, we have relied on Sample Registration Survey (SRS, 2011).

Access to Safe Drinking Water and Sanitation: Regional Trends

Drinking water and sanitation are two basic and important items of social basket that are required by every human being for bare minimum living. Providing safe drinking water and proper sanitation to millions of people is really a big challenge. According to Census 2011, in Rajasthan 82 per cent households have access to improved source of drinking water, whereas only 13 per cent households have improved source of sanitation. This is reflected in state's poor health outcomes, such as infant mortality rate, child mortality rate, communicable diseases and overall morbidity. Rural-urban disparity in access to safe drinking water and improved sanitation is quite evident from 2011 census data. In urban areas, people to a large extent, depend upon water supply systems that either draw from proximate surface water bodies or draw sub-surface water. Water treatment plants are either lacking in many areas or are maintained inadequately. There are very few areas having continuous water supply and many cities have the problem of infrequent and irregular water supply. Poor and slum areas have the sharing points of water supply such as public taps and tube wells which serve the underserved households. In rural areas, shared hand pumps, tube wells and tap water are commonly found as main source of drinking water. Over the past decade, significant improvements have been witnessed but it remains inadequate in terms of overall coverage and good quality, due to rapid urbanization and low resource allocation, among other factors (Mahal et al., 2010).

Table 12.1

Summary Statistics of Population having Access to Safe Drinking Water

	Total		*Rural*		*Urban*		*SC*		*ST*	
	2001	*2011*	*2001*	*2011*	*2001*	*2011*	*2001*	*2011*	*2001*	*2011*
Average	32.62	29.00	20.53	17.71	77.51	70.08	29.79	27.68	21.97	19.97
S.D.	14.60	16.87	13.07	14.61	9.38	11.32	13.14	14.46	20.21	20.70
C.V.	44.77	58.17	63.65	82.53	12.10	15.92	44.11	52.24	92.00	103.67

Note: Authors' own calculation.

Variation in the rate at which regions have increased access to safe water is striking. In Table 12.1, data used to track progress towards the MDGs drinking water target are disaggregated by residence and by caste. With all averages going down and coefficient of variation figures going up, it paints a dismal picture. District-wise variation in Scheduled Tribe population may be explained due to concentration of tribal population in certain pockets but too high variation in total and urban population calls for intervention. The results show stark disparities in coverage, illustrating the challenges in equitable achievement of the MDGs. An estimated 70 per cent of the urban population had access to safe water in 2011, compared to meagre 18 per cent of the rural population. In all, 7.8 million rural households still lack accessibility to tap water. The significant increase in the urban population that gained access to improved water sources between 2001 and 2011 is well over 1.8 million households. However,

Figure 12.1

Water Supply in Rajasthan: By Source and Residence

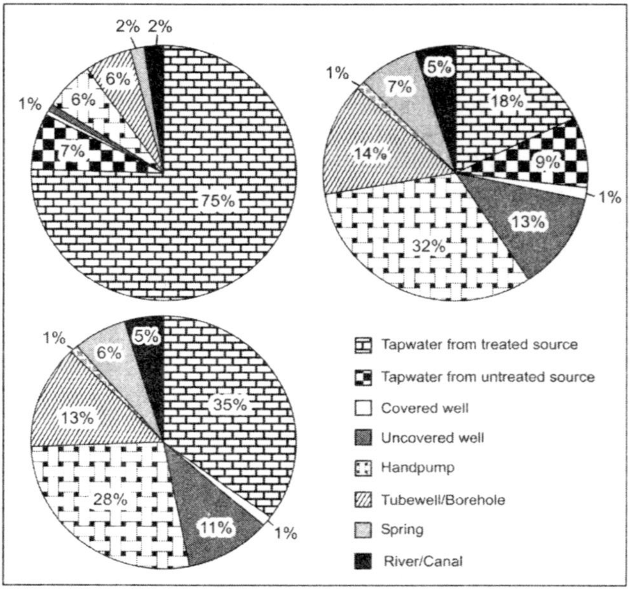

Source: Based on Census, 2011.

Figure 12.2

Accessibility of Tap Water from Treated Source: By Region

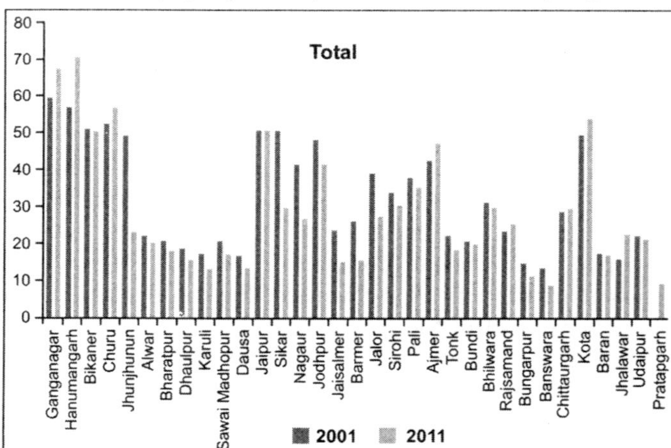

Source: Based on Census, 2011.

the number of urban households using unimproved sources actually increased from 6.04 million to 7.4 million.

This must be viewed in relation to the massive growth in urban population over the same time period. Corresponding figures for rural areas are 5.6 million in 2001 going up to 6.9 million households in 2011. Though the increase in the absolute number of households not having access to safe water over a decade is somewhat parallel in rural and urban areas; the number of households in rural areas using other than tap water source in 2011 was still 13 times greater than in urban areas. The figures also illustrate caste wise disparities; which is not only high but is also not showing any sign of improvement. Especially among Scheduled Tribe too, low average figure merits immediate attention.

So far as sanitation is concerned; people from urban areas mainly depend on sanitation system or septic tanks for disposal. However, it is quite unfortunate that even today, these are inaccessible to many people and as a result, they have been depending on open defecation, especially in poor and slum areas. Drainage system is non-existent in many areas, which results in contamination of surface and subsurface water, making it polluted. In rural areas majority of the people depend on open defecation. Closed drainage systems in rural areas need much more improvement. Besides, lack of awareness, poor habits and hygiene make the situation worse.

Table 12.2

Households not having Access to Improved Sanitation in Rajasthan, 2011 (in percent)

Residence	*No Latrine*	*No Drainage*	*No Bathroom*
Rural	80.35	72.96	55.44
Urban	17.98	13.96	10.82
Total	65.02	58.46	44.48

Note: Authors' own calculation.

Linkages of Drinking Water and Sanitation

Improved water supply and sanitation are recognized as two most important prerequisites for good health and socio-economic development. This has been recognized by the 'Right to Water and Sanitation' adopted by the Committee on Economic, Social and Cultural Rights (CESCR) and the United Nations Sub-Commission on Promotion and Protection of Human Rights in 2002, which includes right to

Figure 12.3

Variations in Closed Drainage

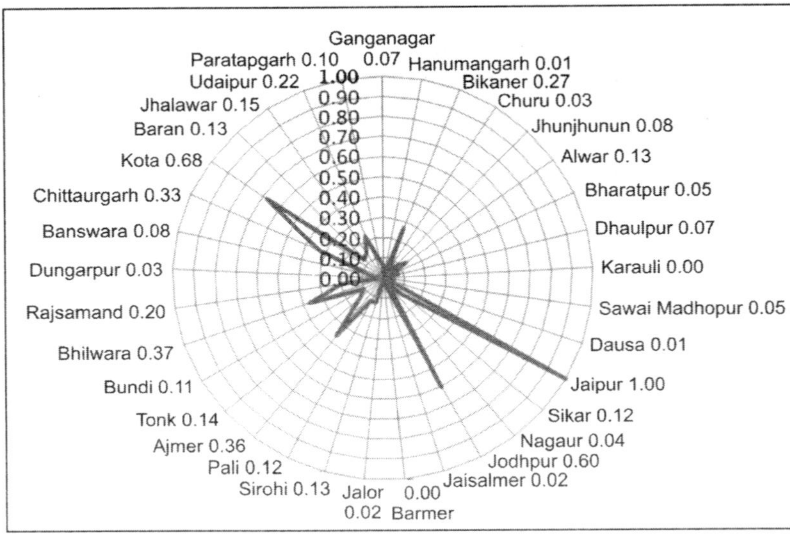

sufficient, clean, accessible and affordable water and sanitation for all. National Health Policy also recognizes the fact that water supply, sanitation and health are very much interrelated with each other. Evidently, both the provisions have forward and backward linkages. Adequate sanitation improves attendance in school, especially among the girls. Health outcomes are positively affected by safe drinking water and improved sanitation. Many other positive externalities in the form of increase in productivity and quality of life too, find place in literature. So, these issues should be addressed holistically. But unfortunately, the reality remains that in India there is very little institutional interface among these departments. The benefits of safe drinking water cannot be realized without proper sanitation system and proper disposal of sewage. In case of open defecation or nonexistence of sewage system, surface and subsurface drinking water tend to be contaminated. In absence of adequate water treatment plant, this unsafe drinking water causes different waterborne diseases. So, it is important to maintain both the quality of water as well as proper sanitation system where sewage is disposed of in an appropriate manner, so that source of drinking water is not contaminated.

In India, diarrhoeal diseases continue to be an important cause of morbidity and mortality in under-fives despite various preventive and standardized case management strategies. It was estimated that in 2005, 3,02,000 children, age 1–59 months, died due to diarrhoeal diseases giving a mortality rate of 11.1 per 1,000 live births (Basani et al., 2010). Nearly three-quarters of child deaths due to diarrhoea occur in just 15 countries of the world and it is sad that India tops the list (WHO, 2008). The transition from unimproved to improved sanitation is accompanied by a more than 30 per cent reduction in child mortality (UNDP, 2006). Economic impact of inadequate sanitation, in addition to health, also includes cost in the form of productivity loss, cost of fetching water, lost tourism earning and cost of treatment of water, etc. The magnitude of loss due to inadequate sanitation in India amounts to ₹ 2.44 trillion (US$ 53.8 billion) a year which was the equivalent of 6.4 per cent of India's GDP in 2006. This means a per person annual impact of ₹ 2180 (US $48). The health related economic impact alone pegged at ₹ 1.75 trillion (US $38.5 billion) accounting for the largest category of impacts. Access time (productive time lost to access sanitation facility) and drinking water related impacts are the other two main losses, at ₹ 487 billion (US $10.7 billion) and ₹ 191 billion (US $4.2 billion) respectively (Tyagi et al., 2006).

Figure 12.4

The Classical Waterborne Infection

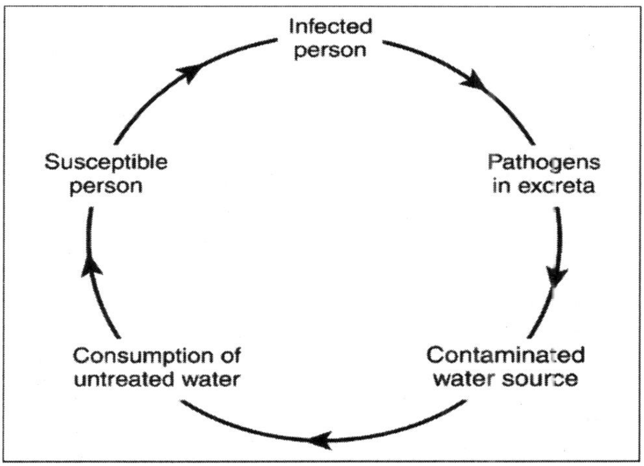

Apart from this interrelationship between drinking water and sanitation, it has also been found that both access to safe drinking water and improved sanitation are influenced by other factors such as per capita income, female literacy, female work participation rate

(FWPR) etc. A number of studies have outlined the significance of income and literacy (especially female education) in improving IMR (Caldwell, 1979, 1986; Cochrane, 1980; Hobecraft et al., 1984; United Nations, 1985; Ware, 1984; Gokhale et al., 2002; Govindasamy and Ramesh, 1997). For this one of the important channels is through access to safe drinking water and improved sanitation. In order to estimate the extent of association among the factors such as per capita income, female literacy and female work participation as well as the parameters that have been taken for the analysis such as percentage of household having no access to improved source of drinking water, percentage of household having no drainage facility, percentage of households having no latrine facility within the premises and percentage of households having no bathroom facility within the premises, Spearman's rank correlation has been calculated. The correlation result is as follows:

A cursory look at the correlation matrix shows that per capita income has a negative and significant correlation with 'no access' to proper source of drinking water as well as drainage facility, latrine facility and bathroom facility. Similar kind of result holds for female literacy and female work participation rates. Female literacy is highly significantly and negatively correlated with no access to drinking water, no drainage facility, no access to latrine facility and no access to bathroom facility implying that improvement in female literacy in the family or society helps in improving access to drinking water and sanitation. Female work participation also positively influences drinking water and sanitation facilities.

All the correlation coefficients are highly significant and also possess expected signs. It is quite evident that access to drinking water and sanitation is positively correlated with Infant Mortality Rate (IMR) and Under-Five Mortality Rate (U5MR) while lack of these facilities bears negative relationship with health outcomes. Statistically high correlation between no access to drinking water and no latrine facility as well as between no access to drinking water and no bathroom facility signifies the interrelationship between drinking water and sanitation.

Route to Better Health Outcomes

Impressive health outcome of a country is concoction of a large number of factors such as adequate and motivated health workers, equitable and accessible health infrastructure, proper water supply and sanitation facilities with hygienic behaviour, enough nutritional provision, especially for the children, etc. All these factors, to some

Table 12.3

Correlation Matrix

	Per Capita Income	*Female Literacy*	*Female WPR*	*No Drinking Water*	*No Drainage*	*No Latrine Facility*	*No Bathroom Facility*	*IMR*	*U5MR*
Per capita Income	–	0.282	–0.574**	–0.534**	–0.482**	–0.565**	–0.429*	–.195	–.138
Female Literacy	–	–	–0.569**	–0.537**	–0.508**	–0.683**	–0.557**	–.577*	–.517**
Female WPR[1]	–	–	–	0.570**	0.399**	0.697**	0.618**	.587**	.475**
No Drinking Water	–	–	–	–	0.421*	0.890**	0.908**	.542**	.530**
No Drainage Facility	–	–	–	–	–	0.467**	0.362*	.479**	.512**
No Latrine Facility	–	–	–	–	–	–	0.860**	.682**	.624**
No Bathroom Facility	–	–	–	–	–	–	–	.587**	.564**
IMR	–	–	–	–	–	–	–	–	.750**
U5MR	–	–	–	–	–	–	–	–	–

** Correlation is significant at the 0.01 level (2 tailed) * Correlation is significant at the 0.05 level (2 tailed)

1. Female Work Pariticpation Rate.

extent, influence health scenario of a country. It has been found that improvements in water supply and sanitation along with effective hygiene education play a major role in reducing high levels of morbidity and mortality in many poor countries (Kurup, 1991; Jalan and Ravallion, 2001; Shreshtha, 2006; World Bank, 2006).

Every year a huge number of children die out of waterborne or water related diseases. To analyse the extent to which unsafe water and improper sanitation influences IMR and U5MR, regression analysis has been carried out.

Table 12.4

Regressions Results

Coefficient (95 % confidence interval)		*P–Value*
Predictor: Proportion of Households without Access to Drainage		
U5MR	0.452 (0.260 – 0.643)	0.000
IMR	0.335(0.176 – 0.494)	0.001
Predictor: Proportion of Households without Access to Latrine Facility		
U5MR	0.236 (0.082 – 0.390)	0.004
IMR	0.201 (0.081 – 0.321)	0.002
Predictor: Proportion of Households without Bathroom Facility		
U5MR	0.256 (0.105 – 0.406)	0.002
IMR	0.198 (0.076 – 0.319)	0.002

Two key outcome variables were explored in this analysis: U5MR and IMR.

U5MR is defined as the probability of a child born in a specific year or period dying before reaching the age of five per 1000 live births IMR is defined as the probability of dying between birth and exactly 1 year of age, expressed per 1000 live births. Data for both the outcome variables were obtained from vital statistics, Sample Registration System, 2011.

Water and sanitation indicators were considered in separate models in the study. Due to considerable overlap between the two variables ($r > 0.85$), water and sanitation access cancelled each other in effect; their simultaneous effect, therefore, was not explored.

In our analyses, increased inaccessibility to sanitation indicators was significantly associated with increased U5MR and IMR. U5MR was seen to increase by 0.452 (95% CI 0.260–0.643) deaths per 1000 with increase in households without drainage, IMR increased by 0.335 (95% CI 0.176 – 0.494) deaths per 1000 with decrease in access to drainage. This increase continued to be significant at $p < 0.001$.

This relationship retained its significance for the predictor variable proportion of households without access to latrine facility. Increase in U5MR and IMR is discernible with decrease in accessibility and these relationships too are significant at $p < 0.005$.

Table 12.4 also presents results for the predictor proportion of households without bathroom facility in the premises. Though the magnitude of increase in U5MR and IMR is a bit low for this, the relationship is highly significant at $p < 0.002$. The R^2 values too stand in between 37 per cent and 45 percent. This allows confidence that the models account for a substantial part of the variability in the outcome measures. A similar model with predictor inaccessibility to safe drinking water did not yield significant results.

The analyses show interesting and statistically significant relationships between sanitation and health. Increasing proportion of households without access to sanitation are significantly associated with increases in the negative health outcomes of interest in the study, namely U5MR and IMR. In the cross section study for all the districts of Rajasthan, we have substantial evidence to accept the strong associations between access to improved sanitation and health. This further reinforces the importance of sanitation, put forward by previous literature.

Conclusion

Recently, the human and social handicaps that inadequate sanitation and unsafe drinking water pose for the people, especially the poor, has been recognized and improving access to drinking water and sanitation is now being given greater priority by the Governments and civil societies. Poor coverage of drinking water and sanitation facility in overall Rajasthan makes the life of the dwellers miserable. Rural sanitation and hygiene in line with Total Sanitation Campaign (TSC) and Government of India (GoI) guidelines is abysmally low in Rajasthan. On an average, only 12.9 per cent people in the state use toilets. According to Nirmal Rajasthan Rural Sanitation and Hygiene Strategy (2012–2022), though access to toilets in schools and Anganwadis has seen a marked increase, open defecation is still widespread (87.5%) in rural Rajasthan, and rural solid and liquid waste management has got little or no attention. Not only are the sanitation conditions poor, water supply is insufficient and excessive fluoride content constitutes an additional health hazard. Recurrent droughts and excessive use of ground water raises the salt content.

Limited financial resources, lack of sustainability of water supply and sanitation services, unhygienic behaviour, and inadequate sanitation in public places such as hospitals, health centres and schools are responsible for the deplorable situation. Lack of proper sanitation facility is a serious health risk and insult to human dignity. By improving source of water a considerable amount of water-borne, water washed, water related diseases can be prevented. Besides treating the water, hygienic water storage is also very important. The transition from unimproved to improved sanitation is accompanied by a more than 30 per cent reduction in child mortality (UNDP, 2006). Recent research also suggests that every US dollar spent on sanitation generates, on average, benefit of 9 US dollars, making it one of the most cost effective development interventions (WHO, 2008). If the State has to come out of red zone of mortality outcomes and has to address the Millennium Development Goal of reducing IMR, the improved water and sanitation accesses are the key strategies.

Unaware of the significant positive externality, individuals generally leave drinking water and sanitation to the State to take care. It's also true that despite wide reaching ramifications, it has been vulnerable to budgetary restraints. Many other priorities crowd the agenda and sanitation is often sidelined. Being root source of poor health, steps to scale up sanitation can help provide better value for money spent on health infrastructure and health delivery system. High rates of population growth and urbanization will further create growing challenges to cope with sanitary provisions for high density, low income populations. Commendable example of Kerala suggests that large gains are possible given the political and institutional structures necessary to convert policy on universalizing access into a reality.

It's better to hit the trunk, branches will fall automatically.

The department of public health and engineering typically focuses on infrastructure creation such as installing hand pumps and tube wells, constructing toilets etc. But only infrastructure creation will not serve the purpose. It is equally important that infrastructure is accessible and usable by all. Much more attention is needed in ensuring quality services, generating public awareness, educating people about hygienic habits and making efficient use of scarce resources.

For wider understanding of the significance of access to water and sanitation in socioeconomic development, sensitization is indispensable. Taboos surrounding the toilet and human excreta hinder progress in this field. Therefore, a stronger appreciation of the manifold societywide benefits of sanitation and the challenges of achieving them is need of the hour. To promote safe hygienic practices at the household and community level and to create sustained behavioural change, innovative strategies must be evolved. For making toilet a desirable household good, the challenge is to offer both the poor and the non-poor a range of desirable and affordable options while persuading them to change their priorities so that improved sanitation becomes an attractive 'must have' for every household.

References

Basani, D.G., Kumar, R., Awasthi, S., Morris, S.K. and Paul, V.K (2010). Causes of Neonatal and Child Mortality in India: A Nationally Representative Mortality Survey, *Lancet*. 376: 1853–60.

Blossner, M. and De Onis, M. (2005). Malnutrition: Quantifying the Health Impact at National and Local Levels, Geneva, World Health Organization. 51.

Caldwell, John C. (1979). Education as Factor in Mortality Decline: An Examination of Nigerian Data, *Population Studies*. 33.

Caldwell, John C. (1986). Routes to Low Mortality in Poor Countries, *Population and Development Review*. 12(2).

Cochrane, S.H. (1980). The Socio-economic Determinants of Mortality: The Cross National Evidence. In Susan H. Cochrane, Donald O. Hara and Joanne Laslie. The Effects of Education on Health, World Bank Staff Working Paper No. 405, Washington DC.

Gokhale, M.K., Rao, S.S. and Garole, V.R. (2002). Infant Mortality in India: Use of Maternal and Child Health Services in Relation to Literacy Status. *Journal of Health Population and Nutrition*. 20(2).

Govindasamy, Pavalavalli and Ramesh, B.M. (1997). Maternal Education and Utilization of Maternal and Child Health Services in India, National Family Heath Survey Subject Report No. 5, Mumbai International Institute for Population Sciences and Calvarton Maryland Macro International.

Hobecraft, J.N., McDonald, R.W. and Rutstein, S.O. (1984). Demographic Determinants of Infant and Early Child Mortality: Comparative Analysis. *Population Studies*.

Hutton, G. and Haller, H. (2004). Evaluation of the Costs and Benefits of Water and Sanitation Improvements at the Global Level. Geneva: World Health Organization, 87.

Jalan, Jyotsna and Martin Ravallion (2001). Does Piped Water Reduce Diarrhoea for Children in Rural India? Indian Statistical Institute and the World Bank (August).

Kurup, K.B. (1991). Social Indicators Research, Vol. 24, No. 4, (June 1991), 403–414.

Mahal Ajay, Debroy, B. and Bhandari, L. (2010). India Health Report, Indicus Analytics, *Business Standard*, New Delhi.

Pruss-Ustun, A., Bos, R., Gore, F. and Bartram, J. (2008). Safer Water, Better Health: Costs, Benefits and Sustainability of Interventions to Protect and Promote Health, Geneva: World Health Organization.

Sample Registration Survey (SRS) (2011). Vital Statistics Division, Office of the Registrar General, India, Ministry of Home Affairs, Government of India.

Shreshtha, Laura B. (2006). Life expectancy in the United States, Congressional Research Service, August.

Tyagi, A., Hutton, G and Sen, S. (2006). The Economic Impacts of Inadequate Sanitation in India, World Bank, Water and Sanitation Program (WSP), USA.

UNDP (2006). Human Development Report: Beyond Scarcity, Power, Poverty and the Global Water Crisis, Oxford University Press.

United Nations (1985). Socio-economic Differentials in Child Mortality in Developing Countries, ST/ESA/SER. A/97, United Nations Publications, New York.

United Nations (2000). United Nations General Assembly, UN Resolution A/55/2.

UNICEF and WHO (2009). Diarrhoea: Why Children are Still Dying and What Can Be Done. The United Nations Children's Fund (UNICEF)/World Health Organization (WHO).

UNICEF and WHO (2012). Joint Monitoring Programme for Water Supply and Sanitation: Progress on Drinking Water and Sanitation. *Update*, New York and Geneva.

Ware, Helen. (1984). Effects of Maternal Education, Women's Roles and Child Care on Child Mortality in Henry W. Hosley and Linclon C. Chen (eds) Child Survival Strategies, Population and Development Review. 10.

WHO (2008). Safer Water, Better Health: Costs, Benefits and Sustainability of Interventions to Protect and Promote Health, World Health Organization, Geneva.

World Bank (2006). India Inclusive Growth and Service Delivery: Building on India's Success, Development Policy Review. Available at http://siteresources.worldbank.org/SOUTHASIAEXT/Resources/DPR_FullReport.pdf.

World Bank (2008). Environmental Health and Child Survival: Epidemiology, Economics Experience, Washington, DC World Bank. 135.

13

Water, Sanitation and Health

An Inter-District Analysis of Rajasthan

Alpana Kateja and Chitra Choudhary

Access to safe drinking water and basic sanitation facilities is recognized as a basic human need and a precondition for economic and social development. Improved water supply and sanitation has been targeted and promoted as essential public health measures for the health of population. The United Nations (2000) confirmed the central role of water and sanitation by including it into one of the Millennium Development Goals (MDG)[1]. WHO (2004) states them as the primary drivers of public health and also imperatives to ensure economic benefits. Significance of water and sanitation has also been reiterated by the United Nations General Assembly Resolution 2010 for realization of all human rights. The WHO and UNICEF (2015) Report states that the target for safe drinking water was met in 2010; over 90 per cent of the world's population has access to improved sources of drinking water; however, the world has fallen short on the sanitation target, leaving 2.4 billion without access to improved sanitation facilities. Huge regional, urban and rural disparities too exist worldwide.

WHO (2008) estimated that meeting the MDG 7 would avert 4,70,000 deaths per year and enhance people's economic livelihood opportunities. Access to safe drinking water and adequate sanitation services are vital to human health and also have other important benefits ranging from the easily identifiable and quantifiable (costs avoided, time saved) to the more intangible (convenience, well-being, dignity, privacy and safety). *The Asia Water Watch* (2015) report estimates that for every US$ 1 invested in people gaining access to water and sanitation, there will be returns of US$ 6 in health, livelihood and education benefits; also an investment of just $8 billion annually will return $54 billion annually which is seven times of the investment. From this perspective, the health and socio-economic benefits of improved access to safe water and adequate sanitation are the most compelling arguments to support resource allocations towards the MDG 7.

The role of access to safe drinking water and sanitation in influencing health outcomes in developing countries has been well understood. Health impact of water and sanitation interventions have been focused and analysed by several studies (Jamie et al., 2005; Montgomery and MeNachem, 2007). A multifaceted review done by Esrey et al. (1991) found significant reductions in both the severity and prevalence of diarrhoea and infectious diseases as a result of improved water supply and sanitation. Cairncross et al. (2010) have demonstrated that the diarrhoeal risk reductions with handwashing using soap, improved water quality and excreta disposal is 48, 17 and 36 per cent respectively. Similar findings have been made by Pruss et al. (2014) regarding the effects of inadequate water and sanitation in low and middle income countries. A number of studies have shown access to improved water and sanitation as the important determinants of child mortality (Kamiya, 2010; Farahani et al., 2010 and Sahoo, 2010). Nearly 60 per cent of infant mortality is linked to infectious diseases, most of them are water, sanitation and hygiene related (UNICEF, 2006). Braghetta (2006) corroborated that lack of safe drinking water contributes significantly to the high incidence of diarrhoeal disease and chronic malnutrition. Manda (2009) also concluded that substantial reduction in the use of unsafe water and sanitation in Malawai would have considerable impact on levels of disease.

Water and sanitation access are not only important from health point of view, but also directly affect all other seven MDGs, which cover wide ranging issues such as, education, gender equality, child mortality, maternal health and infectious diseases, poverty and hunger (UNDP, 2005). Thus, MDG 7 is instrumental in achieving all MDGs. Another study by Kumar and Das (2014) examine the level, trend and progress rate of MDG 7 in India using the census data and also investigate their association with diarrhoea. They concluded that India is on track to meet the target on access to safe drinking water; however, it is lagging far behind in terms of sanitation access. Water and sanitation are at the forefront of the international debates on sustainable development and to accelerate the efforts to achieve universal sanitation coverage, Government of India has launched the Swachh Bharat Mission on 2 October 2014. Rajasthan has also made concerted efforts for the well-being of its population by expanding access to water and sanitation. State is still far behind from achieving goal of MDG 7 and has to travel a long way for the same. The present study provides an analysis of the progress in access and availability of safe water and improved sanitation up to the period 2011 in all the districts of Rajasthan. Though significant progress has been reported in water access, a majority of the population do not have access to sanitation and resort to open defecation. According to Census 2011, about 71.88 per cent of the households are reported to have connected with tap water, over 53.16 per cent and nearly 87.32 per cent is for rural and urban population respectively. However, only 34 per cent have latrine facility within the premises; large rural-urban variations exist as only 19.64 per cent rural and 82 per cent urban population have toilet facilities. Around 64.32 per cent of population is practising open defecation in the state. This silent humanitarian crisis thwarts socio-economic progress.

Following are the objectives of the study:

(1) To provide a broad overview of access to safe drinking water and sanitation facilities in all the districts of Rajasthan.
(2) To provide critical inter-district insights into rural-urban differentials with respect to these basic amenities.
(3) To investigate the performance of all districts of Rajasthan in ensuring attainment of MDG 7.

Data Sources and Methodology

The study attempts to unravel the inter-district and rural-urban variations in Rajasthan state in terms of accessibility and availability of safe water and improved sanitation over the period of twenty years i.e. from 1991 to 2011. Data on safe drinking water and sanitation are collected from houses, household amenities and assets tables of Census of India, 2011 and Human Development Report (Rajasthan, 2008). Applying rotated PCA (Principal Component Analysis), an attempt has been made to construct a comprehensive measure of Water-Sanitation Index for rural and urban areas using four variables namely; (A) *Access to safe drinking water*: (i) It is the percentage of households who have access to safe drinking water (viz. tap water from treated sources) within the premises. (B) *Access to sanitation*: It is the percentage of households who have access to improved sanitation facilities within the premises viz. (ii) latrine, (iii) bathroom and (iv) closed drainage facilities.

Principal component analysis (PCA) is a statistical procedure that uses an orthogonal transformation to convert a set of observations of possibly correlated variables into a set of values of linearly uncorrelated variables called principal components. This transformation is defined in such a way that the first principal component has the largest possible variance (that is, accounts for as much of the variability in the data as possible) and each succeeding component in turn has the highest variance possible under the constraint that it is orthogonal to the preceding components. The principal components are orthogonal because they are the eigen vectors of the covariance matrix, which is symmetric. From an initial set of n correlated variables, PCA creates uncorrelated indices or components, where each component is a linear weighted combination of the initial variables. For example, for a set of variables X_1 to X_n,

$$PC_1 = a_{11}X_1 + a_{12}X_2 + \ldots + a_{1n}X_n$$
$$PC_2 = a_{21}X + a_{22}X_2 + \ldots + a_{2n}X_n.$$
$$\ldots\ldots$$
$$\ldots\ldots$$
$$\ldots\ldots$$
$$PC_m = a_{m1}X_1 + a_{m2}X_2 + \ldots + a_{mn}X_n$$

Where a_{mn} represents the weight for the mth principal component and the nth variable.

After constructing water-sanitation index for rural and urban area, districts have been classified into High, Medium and Low category. In order to examine the progress towards achieving MDG 7 i.e. to halve the number of people who do not have sustainable access to safe drinking water and basic sanitation by 2015, following methodology proposed by Kumar and Das (2014) has been used.

Table 13.1

Methodology of Computing the Annual Actual Rate of Progress (AARP) and Required Rate of Annual Progress (RRAP)

Indicator	*Target*	*AARP*	*RRAP*
Basic Sanitation Facility Access to Safe Drinking Water	Reduce by half the proportion of people without sustainable access	$\frac{(X_{t1} - X_{t0}) * 100}{X_{10} * (X_{t1} - X_{t0})}$	$\frac{(X_{tmdg} - X_{t0}) * 100}{X_{10} * (X_{tmdg} - X_{t0})}$

Where, t_1 refers to period of 2011; X_{t1} are the values of the 2011 year for both the variables access to safe drinking water and sanitation, X_{t0} is the value of indicator for the period 1991; X_{tmdg} is the value of indicator which must be achieved in 2015 according to MDG, lastly in the end X_{tmdg} is the year (2015) by which the target is to be met and t_0 is the year 1991.

According to UNDP, *Improved water sources as identified for MDG target include*: (a) piped water into dwelling, plot/land; (b) public tap/stand pipe; (c) tubewell/borewell; (d) protected dugwell; (e) protected spring; and (f) rainwater collection/harvested rainwater. *Improved sanitation facilities include*: (a) flush/pour flush into septic tank, piped sewer system, or pit; (b) ventilated improved pit latrine; (c) pit latrine with slab; and (d) compositing toilet. The study is confined to access to tap water from treated sources as it is a major source of access to water. However, performance of all the districts towards access to water is relatively much better than sanitation. The remainder of the paper, highlights intra-state and rural-urban variations with respect to these basic amenities. Then, on the basis of constructed rural-urban Water Sanitation Index, all the districts have been ranked and classified. The last section explains the progress of all districts towards accessing water and sanitation which ensures progress of the Rajasthan state towards achieving its goal of MDG 7 in general.

Access to Safe Drinking Water and Sanitation in Rajasthan

Drinking water and sanitation are crucial inputs for improved health outcomes. The provision of improved sanitation and safe drinking water could reduce diarrhoeal diseases by nearly 90 per cent (UNICEF, 2006). Improved water supply and sanitation reduce diarrhoea morbidity by 21 and 37.5 per cent respectively (WHO, 2011). According to Census 2011, in Rajasthan state 71.83 per cent have access to safe drinking water; in rural areas only 53.16 per cent have drinking water facility within their premises, in comparison to their urban counterparts (87.32%). Large inter-district variations are evident. In Ajmer, 94 per cent households have access to tap water followed by Jodhpur (88%), Hanumangarh (89%), Ganganagar (87%), Bikaner (85%)and Bhilwara (82%) while in some of the districts such as Banswara, Dungarpur, Karauli, not even 50 per cent household have access to safe drinking water. Figure 13.1 depicts scenario of huge rural-urban differential among districts in terms of 'availability of Tap water'. Rural areas of almost all the districts are lagging far behind in comparison to their urban counterparts. Stark disparities on the basis of residence are confirmed by 't statistic' also. On an average, in rural areas only 49 per cent have access to tap water relative to 84.16 per cent in urban areas of all the districts (Table 13.2).

Figure 13.1

Availability of Tap Treated Water (Within the Premises)

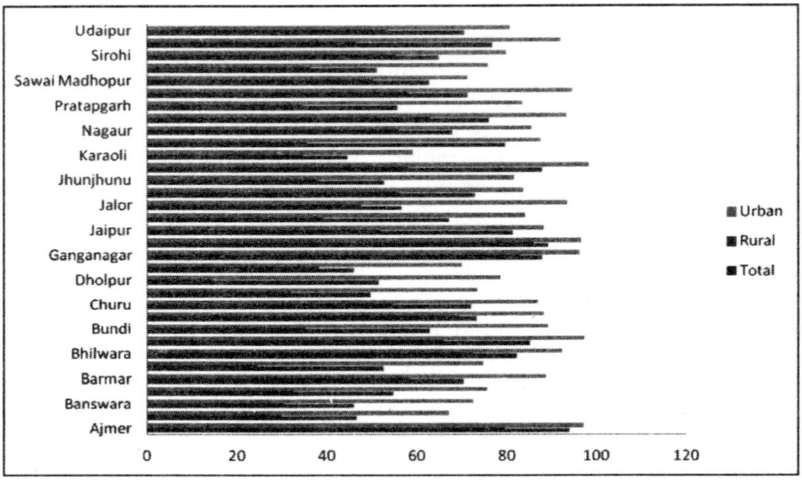

Source: Based on Census 2011 data.

Table 13.2

Descriptive Statistics: Households having Access to Safe Drinking Water

2011 (N=33)	*Total*	*Rural*	*Urban*
Average	66.7067	49.071	84.161
S.D.	14.54871	17.404	9.97056
C.V.	21.81	35.47	11.85
R-U Differences t value (–32.267)* 95% confidence interval[–47.680 _ (–32.4090)]			

Note: *refers to 1 per cent level of significance.

Access to sanitation facility: Sanitation has a greater impact on child health, based on mortality, growth and morbidity indicators. It involves both behaviour and facilities, which can together form a hygienic environment. The practice of open defecation comes from a combination of factors: the most prominent of them being the traditional behavioural pattern and lack of awareness of the people about the associated health hazards. Large inter-district variations are reflected in Figure 13.2.

Figure 13.2

Availability of Bathing Facility (Within the Premises)

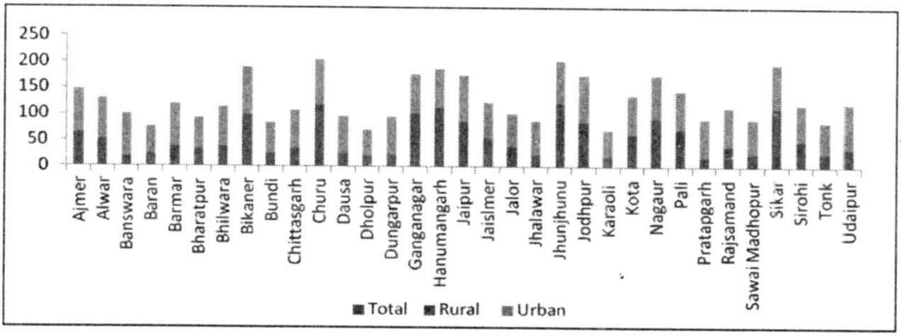

Source: Based on Census 2011 data.

Table 13.3

Descriptive Statistics: Households having Access to Bathing Facility

2011 (N=33)	*Total*	*Rural*	*Urban*
Average	32.5821	21.533	73.182
S.D.	18.03653	16.730	11.087
C.V.	55.36	77.69	15.15
R-U Differences t value (–29.523)* 95% confidence interval [–54.650 _ (–49.648)]			

Note: *refers to 1 per cent level of significance.

In Churu, Jhunhunu, Jaipur, Sikar, Hanumangarh, Ganganagar, Nagaur and Jodhpur districts, households have access to bathing facility more than the average (32.58%). While Banswara, Karauli, Pratapgarh, Dungarpur, Bundi and Dholpur have poor sanitation conditions (10 to 15%). In rural areas, very few districts are above average. Urban areas have shown better sanitation practices than rural areas in all the districts.

Table 13.4

Descriptive Statistics: Households having Access to Latrine Facility

2011 (N=33)	*Total*	*Rural*	*Urban*
Average	31.6361	19.1982	76.867
S.D.	20.20394	19.5079	10.762
C.V.	63.86	101.61	14.00
R-U Differences t value (–35.375)* 95% confidence interval [–60.770 _ (–54.5681)]			

Note: *refers to 1 per cent level of significance.

Figure 13.3

Availability of Latrine Facility (Within Premises)

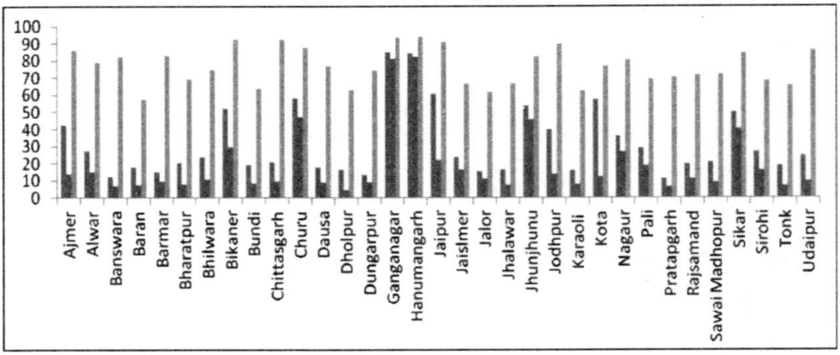

Source: Based on Census 2011 data.

On an average, 31.63 per cent households have access to latrine facility. Jaipur, Hanumangarh and Ganganagar are in better position (Figure 13.3). While in Banswara, Barmer, Bundi, Baran, Dausa, Dholpur, Jhalawar, Karauli, Tonk districts, not even 20 per cent population have access to it. In rural areas, only Sikar, Nagaur, Jhunjhunu, Ganganagar, Hanumangarh and Bikaner districts more population has access than the state average of 19 per cent.

Drainage system is non-existent in many districts (Figure 13.4), which results in contamination of surface and sub-surface water.

Figure 13.4

Inter-District Variation in Closed Drainage, 2011

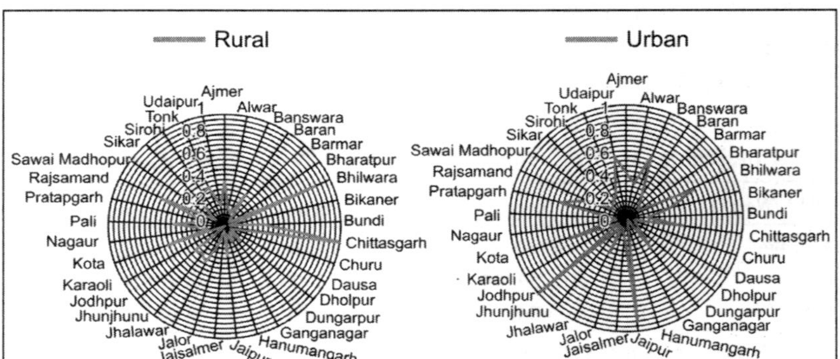

Source: Based on Census 2011 data.

Table 13.5

Descriptive Statistics: Households with Access to Drainage Facility

2011 (N=33)	*Total*	*Rural*	*Urban*
Average	8.2055	2.969	23.930
S.D.	7.58571	1.734	15.340
C.V.	92.45	58.40	64.10
R-U Differences t value (.877)* 95% confidence interval [–16.1362 _ (–15.6856)]			

Note: *refers to 1 per cent level of significance.

On an average only 8.21 per cent people have closed drainage facility within the premises; rural areas have merely 2.96 per cent as compared to urban counterparts 23.93 per cent, which in itself is modest.

A major proportion of households, especially in rural areas, lack in terms of availability of drinking water and sanitation facility within the premises. Women have to travel a long distance to get water for drinking, cooking and other household chores. During pregnancy, collecting water from a far distance becomes hazardous for the mother as well as for child's health. So, it is essential to step up actions to improve the profile of the availability of drinking water, especially at village level considering maternal and child health. People in urban areas are far more likely to have both the facilities. The difference is disproportionately large in worst performing districts such as Banswara, Bundi, Barmer and Bharatpur. To analyse the structural differences in the accessibility of safe drinking water and sanitation across rural and urban areas, t-test for differences in means has been employed. The rural-urban differences have been found to be highly significant for both water and sanitation provision. Moreover, rural-urban divide is more visible in case of access to sanitary facilities than the drinking water.

Water-Sanitation Index for Rajasthan State

Access to safe drinking water, latrine, bathroom and closed drainage facilities together represent the water and sanitation provision in the analysis. Water-Sanitation Index has been constructed by using the technique of Principal Component Analysis (PCA). For rural and urban areas, this index is a linear combination of weighted variables explaining 55 and 58.90 per cent of total variance. With Kaiser-Meyer-Olkin (KMO) measure of sampling adequacy being equal to 0.533 and 0.621 and Bartlett's Test of Spherecity being equal to 56.233 and 50.973 (p value .000) respectively.

Table 13.6

Water-Sanitation Index and Ranking of Districts

Districts	*Factor loadings (Rural)*	*Rank*	*Factor loadings (Urban)*	*Rank*
Ajmer	0.2069	11 (M)	1.1689	4 (H)
Alwar	–0.3705	16 (M)	–0.2004	19 (M)
Banswara	–0.8725	31 (L)	0.4961	12 (M)
Baran	–0.9027	32 (L)	–1.7214	32 (L)
Barmer	–0.1000	20 (M)	0.4964	11 (M)
Bharatpur	–0.8150	28 (L)	–1.0638	30 (L)
Bhilwara	–0.3812	18 (M)	0.5412	10 (H)
Bikaner	1.0640	5 (H)	1.5963	3 (H)
Bundi	–0.8557	30 (L)	–0.9151	28 (L)
Chittorgarh	–0.5470	21 (M)	0.8052	6 (H)
Churu	1.6614	3 (H)	0.6695	9 (H)
Dausa	–0.7170	25 (L)	–0.5449	22 (L)
Dholpur	–1.2131	33 (L)	–1.5589	31 (L)
Dungarpur	–0.598	23 (L)	–0.4726	21 (M)
Ganganagar	2.6032	2 (H)	0.7879	7 (H)
Hanumangarh	2.8809	1 (H)	0.6837	8 (H)
Jaipur	0.2106	10 (H)	1.7587	2 (H)
Jaisalmer	–0.0059	12 (M)	–0.7617	26 (L)
Jalore	–0.2487	15 (M)	–0.8737	27 (L)
Jhalawar	–0.5603	22 (L)	–0.6543	23 (L)
Jhunjhunu	1.3158	4 (H)	0.2395	16 (M)
Jodhpur	0.2755	9 (H)	2.1392	1 (H)
Karauli	–0.7105	24 (L)	–2.1532	33 (L)
Kota	–0.7392	27 (L)	0.3740	13 (M)
Nagaur	0.8811	7 (H)	0.2454	15 (M)
Pali	0.3804	8 (H)	–0.2217	20 (M)
Pratapgarh	–0.8431	29 (L)	0.0205	18 (M)
Rajsamand	–0.3775	17 (M)	0.1821	17 (M)
Sawai Madhopur	–0.4887	20 (M)	–0.9422	29 (L)
Sikar	1.0434	6 (H)	0.3366	14 (M)
Sirohi	–0.0420	13 (M)	–0.6711	24 (L)
Tonk	–0.7304	26 (L)	–0.6990	25 (L)
Udaipur	–0.4038	19 (M)	0.9125	5 (H)

Water-Sanitation Index for rural and urban population (Table 13.6) shows accessibility of water and sanitation facilities at district level. It is quite evident that with respect to accessibility of these services for rural population, Hanumangarh has the highest rank followed by Ganganagar, Churu, Jhunjhunu, Bikaner, Sikar, Nagaur, Pali, Jodhpur and Jaipur while Dholpur has worst performance followed by Banswara, Baran, Bundi, Dausa, Dungarpur, Bharatpur, Karauli, Kota and Pratapgarh respectively. However, mediocre districts are Barmer, Bhilwara, Chattisgarh, Jaisalmer, Jalore and Udaipur respectively. The only district with diversified performance is Kota with low score of index (27). On the urban counterparts Jodhpur, ranks first followed by Jaipur, Bikaner, Ajmer, Udaipur, Chattisgarh, Ganganagar and Hanumangarh vis-à-vis mediocre districts are Banswara, Barmer, Kota, Sikar and Nagaur. Karauli has the worst picture followed by Dholpur, Baran and Bharatpur respectively. The analysis concludes that there exist inter-district high disparity between rural and urban population in terms of access to improved water and sanitation.

Sustainable access to clean water and proper sanitation are critical for the well-being of all. Table 13.7 reflects that all the districts have performed relatively better in terms of accessibility of improved water rather than sanitation. Given the 1990 level for households without access to improved water and sanitation facility, Rajasthan is required to reduce the proportion of households having no access to improved water and sanitation by 2015. District-wise variations are there up to a large extent. The analysis exhibits that target for access to improved drinking water is not so far while that for improved sanitation is distant.

Table 13.7

Actual (1990–2011) and Required Annual Rate of Progress with Respect to Improved Water and Sanitation (District-wise)

Districts	*AARP**	*RARP**	*AARP***	*RARP***
Ajmer	1.4470	0.7739	1.7521	04.4711
Alwar	–0.1989	2.1152	5.2308	13.2671
Banswara	–1.0788	1.3613	2.1317	21.2558
Baran	–1.3274	0.6385	–1.554	05.6628
Barmer	4.2882	3.4054	4.9380	25.3972
Bharatpur	4.8608	5.6923	2.7827	13.6250

Cont'd...

...Cont'd

Bhilwara	1.7278	1.3167	4.1005	13.8730
Bikaner	2.1016	1.3898	2.3837	03.7803
Bundi	-0.5846	0.7932	2.7010	14.3934
Chittorgarh	0.0645	0.7662	4.1484	16.0180
Churu	0.9381	1.3277	12.998	10.9032
Dausa	-1.2119	1.0120	-2.3305	03.8823
Dholpur	1.6250	3.2219	2.4735	16.5185
Dungarpur	-1.4213	1.0487	4.4996	26.9855
Ganganagar	2.1578	1.3112	2.8365	01.7664
Hanumangarh	2.2579	1.3112	2.8293	01.7664
Jaipur	1.0692	1.0120	3.7212	03.8823
Jaisalmer	0.2639	1.1496	4.3442	14.3934
Jalore	0.0780	1.6036	7.5555	31.3333
Jhalawar	1.9203	1.8535	2.0146	15.0940
Jhunjhunu	-0.5902	1.3277	11.698	10.9032
Jodhpur	1.3657	0.9325	2.1866	05.3529
Karauli	0.8217	3.2770	2.0592	16.0180
Kota	0.2399	0.6385	5.7124	05.6628
Nagaur	1.8172	0.1959	7.5162	03.0971
Pali	1.2945	1.3500	6.2780	14.0000
Rajsamand	0.5229	1.1152	2.0442	12.2857
Sawai Madhopur	3.1109	3.2770	4.1012	16.0180
Sikar	-0.8730	1.2051	9.3322	09.8343
Sirohi	-0.5913	0.7027	5.2120	13.6250
Tonk	1.4806	1.4188	2.2347	13.7480
Udaipur	0.4739	1.1152	3.5340	12.2857

Note: * Reflects actual and required annual rates of access to improved water.
** Reflects actual and required annual rates of access to improved sanitation.

From the above table, it is clear that some districts (Ajmer, Bhilwara, Bikaner, Ganganagar, Hanumangarh, Jhalawar, Jodhpur, Nagaur, Tonk and Barmer) have higher actual rate of improved water than their required rate; while five districts (Sawai Madhopur, Pali, Kota, Churu and Bharatpur) are on track to meet the required rate. Rest sixteen districts are lagging behind in terms of improved water but the difference between actual and required rate is not so much. Meeting the challenge of improved sanitation is more urgent and remains distant. Only Churu, Ganganagar, Hanumangarh, Jhunjhunu, Kota and Nagaur have achieved their required rates although only two districts, Sikar and Jaipur, are on the way to meet the targeted rate.

Remaining twenty-four districts are far behind their required annual rate of progress. If Rajasthan has to come out of red zone of EAG (Empowered Action Group) states, equitable allocation of resources along with strategic plans required to have balanced development of all the districts are urgently required.

Conclusion

The social and economic costs of water and sanitation related diseases are well established. Typical consequences are loss of income and human productivity, resulting in more poverty and loss of gross national product (GNP). It has been calculated that India loses 180 crore man-hours each year due to diseases related to water and sanitation (UNICEF, 2000). Total loss for the rural households in the poorest quintile amount to ₹ 204 billion, which are substantial compared to their urban counterparts (₹ 16 billion). It is quite evident that the problem of inadequate sanitation is a serious concern in rural areas. WHO (2008) estimates that comprehensive sanitation and hygiene interventions can avert 45 per cent of health related economic losses and translate into a potential gain of about ₹ 1.5 trillion, ₹ 1,321 in per capita terms. Adequate and safe water supply is necessary for good sanitation and hygiene, and to underpin livelihoods, nutrition and economic growth. Despite the fact that there exists a close connection between poor sanitary conditions and contaminated water and severe health hazards, access to improved sanitation continues to be an unmet challenge. The study reflects that although the state is about to achieve its target of improved water but it is off the track to meet the target of improved sanitation with vast rural-urban disparities. Economic development makes no sense if majority of the population lives in unhealthy and unclean surroundings due to lack of access to safe water and sanitation. Sanitation involves both behaviours and facilities, which can together form a hygienic environment. Similarly, access to safe drinking water reduces the incidence of illness and water related diseases. Thus, poor water and sanitation facilities have many other serious repercussions. Inadequate and inefficient delivery of these basic amenities in the rural areas put enormous burden of diseases on rural households. Huge amount of out-of-pocket expenditure on medical treatments by the rural households can be saved if all the households have access to uncontaminated drinking water and sanitary toilets properly linked with the sewage system.

The benefits of safe drinking water cannot be realized without proper sanitation system and proper disposal of sewage. In case of open defecation or non-existence of sewage system, surface and subsurface drinking water tend to be contaminated. In absence of adequate water treatment plant this unsafe drinking water causes different waterborne diseases. So, it is important to maintain both the quality of water as well as proper sanitation system where sewage is disposed off in an appropriate manner, so that sources of drinking water are not contaminated. Evidently, both the provisions have forward and backward linkages. Inadequate water and sanitation facility in the community is adding to a lot of burden to the members of the community and bad health, so there is an urgent need to step up some innovative actions for the sustainable development of the community. An integrated approach linking water, sanitation, hygiene and health would improve the quality of living of the entire community. A plan for community based management and maintenance of water, waste and sanitation facilities would ensure that the entire community and not just a handful of people derive benefits, thereby assuring equity and sustainability.

Note

1. MDG Goal 7 which is concerned mainly with environmental sustainability is to halve the number of people who do not have sustainable access to safe drinking water and basic sanitation by 2015.

References

Asia Water Watch (2015). Are Countries in Asia on Track to Meet Target 10 of the MDG's?. Report Published by the Asian Development Bank (ADB), United Nations Development Program (UNDP), World Health Organization (WHO), United Nations Economic and Social Commission for Asia and the Pacific (UNESCAP).

Braghetta, Anne (2006). Drawing the Connection between Malnutrition and Lack of Safe Drinking Water in Guatimala. *Journal of American Water Works Association*, 98(5): 97–106.

Cairncross, Sandy, Caroline Hunt, Sophie Boisson, Kristof Bostoen, Val Curtis, Isaac C.H. Fung and Wolf Peter Schmidt (2010). Water, Sanitation and Hygiene for the Prevention of Diarrhoea. *International Journal of Epidemiology*, 39: 193–205.

Census of India (2011). Office of the Registrar General of India and Census Commissioner. Government of India, New Delhi.

Esrey, S.A., J.B. Potash, L. Roberts and C. Shiff (1991). Effects of Improved Water Supply and Sanitation on Ascariasis, Diarrhoea, Dracunciliasis, Hookworm Infection, Schistosomiasis and Trachoma. Bulletin of World Health Organisation. 69(5), 609–621.

Farahani, M, Subramanian, S.V. and Canning, D. (2010). Effects of State-level Public Spending on Health on the Mortality Probability in India. *Health Economics*, 19 Nov. 11. 1361–1376.

HDR (2008). Human Development Report Rajasthan (An Update), Institute of Development Studies and Government of Rajasthan.

Jamie, Bartram, Kristen, L., Roberto, L. and Wright, A. (2005). Focusing on Improved Water and Sanitation for Health. *The Lancet*, 365(9461), 26th February. 810–812.

Kamiya, Yusuke (2010). Determinants of Health in Developing Countries: Cross-Country Evidence. Osaka School of International Public Policy (OSIPP) Discussion Paper, 15 December, DP–2010, E–009.

Kumar, A. and Das, K.C. (2014). Drinking Water and Sanitation Facility in India and Its Linkages with Diarrhoea among Children Under Five: Evidences from Recent Data. *International Journal of Humanities and Social Science Invention*. 3(4): 50–60.

Manda, Mtafu A. Zeleza (2009). Water and Sanitation in Urban Malawai: Can the MDG's Be Met? A Study of Informal Settlements in Five Cities. Human Settlements Working Paper, Aug., International Institute for Environment and Development (IIED).

Montgomery, Maggie A., and MeNachem Elimelech (2007). Water and Sanitation in Developing Countries: Including Health in the Equation. *Journal of Environmental Science and Technology*, January. 17–24.

Pruss-Ustun, Annette, Jamie Bartram, Thomas Clasen, John M. Colford, Oliver Cumming, Valerie Curtis, Sophie Bonjour, Alan D. Dangour, Jennifer De France, Lorna Fewtrell, Matthew C. Freeman, Bruce Gordon, Paul R. Hunter, Richard B. Johnston, Colin Mathers, Daniel Mausezahi, Kate Medlicott, Maria Neira, Meredith Stocks, Jennyfer Wolf and Sandy Cairncross (2014). Burden of Disease from Inadequate Water, Sanitation and Hygiene in Low and Middle Income Settings: A Retrospective Analysis of Data from 145 Countries. *Tropical Medicine and International Health*, 19(8), August. 894–905.

Sahoo, Dukhabandhu (2010). Dynamics of Health Status and Health Care in Assam, in Himanshu Sekhar Rout and S. Bhyrava Murthy (ed.), *Human Development in India: Challenges and Policies*, July (New Delhi: New Century Publications) India, 158–186.

UN (2000): United Nations Millennium Declaration, 18th Sep., United Nations.

UN General Assembly Resolution (2010). The Human Right to Water and Sanitation, July-A/64/L.63/Rev.1, http://www.unesco.org/water/wwap/news/archives/UNDecWaterHR_EN.pdf.

UNDP (2005). UN Millennium Project Taskforce on Water and Sanitation: Health, Dignity and Development; What will it Take?, United Nations Development Programme, London.

UNICEF (2000). Learning from Experience: Water and Environmental Sanitation in India, United Nations Children's Fund, New York.

UNICEF (2006). Progress for Children: A Report Card on Water and Sanitation, United Nations Children's Fund, New York.

United Nations (2000). United Nations Millennium Declaration, United Nations General Assembly, New York.

WHO (2004). Meeting the MDG Drinking Water and Sanitation Targets: A Mid-term Assessment of Progress. World Health Organization and United Nations Children's Fund, Geneva and New York.

WHO (2008). Safer Water, Better Health: Costs, Benefits and Sustainability of Interventions to Protect and Promote Health. World Health Organization, Geneva.

WHO (2011). Unsafe Water, Sanitation and Hygiene Links to Health. World Health Organization, Geneva.

WHO and UNICEF (2012). Progress on Drinking Water and Sanitation, Joint Monitoring Programme for Water Supply and Sanitation, United Nations Children's Fund. New York and World Health Organisation, Geneva.

WHO and UNICEF (2015). Progress on Drinking Water and Sanitation 2015 Update and MDG Assessment, United Nations Children's Fund. New York and World Health Organisation, Geneva.

14

Ground Water Scenario of Rajasthan

P.K. Parchure

Rajasthan, the largest state of India, with arid and semi-arid climate, erratic rainfall, unique geological conditions, frequent droughts combined with various anthropogenic activities pose severe threat to already fragile condition of water. Since surface water is scarce, demand of water is increasing exponentially; the dependency on ground water has been ever increasing, resulting in alarming overdraft and shrinking safe zones. The present paper makes an effort to obtain information on ground water levels of Rajasthan through representative sampling done by Central Ground Water Board. The most important attribute of ground water regime monitoring is ground water level. The natural conditions affecting the regime involve climatic parameters like rainfall, evapotranspiration etc. whereas anthropogenic influences include withdrawal from the aquifer, recharge due to irrigation systems and other practices like waste disposal etc. To measure the ground water levels, samples were collected from 366 observation wells once a year during the month of April/May to obtain background information of ground water quality changes. The data is used for assessment of ground water resources and changes in the regime consequent to various development and management activities. The paper discusses changes in water parameters namely Electrical Conductivity (EC), Fluoride and Nitrate which have enormous impact on human health.

Physiography

The state has a fairly mature topography developed during the long period of denudation and erosion. Physiographically the State can be divided into four units Aravalli Hill Ranges, Eastern plains, Western Sandy Plain with Sand Dunes and Vindhyan Scarpland and Deccan Lava Plateau (Figure 14.1):

Figure 14.1

Map showing Physiography Rajasthan.

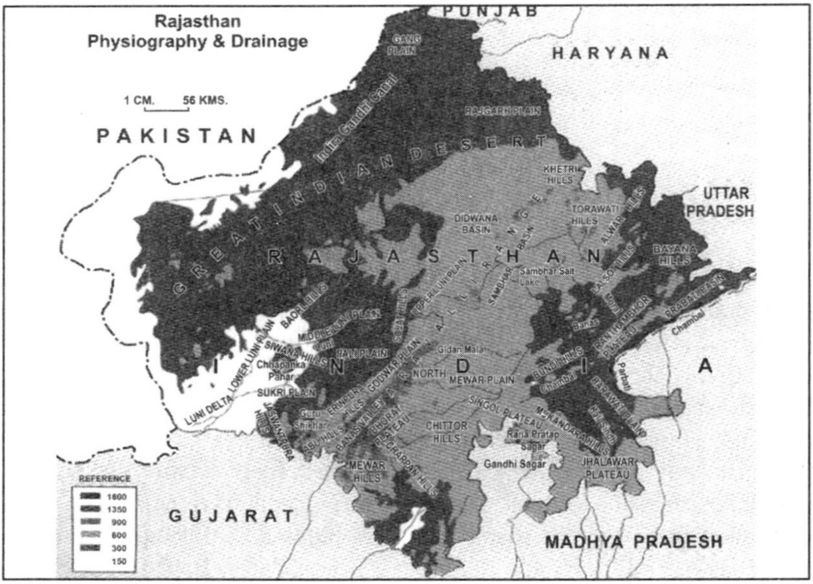

The Western Sandy Plains and Sand Dunes

The sandy plains in western Rajasthan, forming a part of Thar Desert, are mainly occupied by alluvium and blown sands. These plains are further subdivided into three units, i.e. Sandy Arid Plain (Marusthali), Semi-arid Transitional Plain and Ghaggar Plain.

The Sandy Arid Plain

It is a typical desert terrain. The line dividing the Sandy Arid Plain and the Semi-arid Transitional Plain as well as Ghaggar Plain is based on climatic parameters and water resource availability.

The eastern boundary of the Semi-arid Transitional Plain is marked by the foothills and their extension on the western side of

Aravalli Ranges. Sand dunes are prominent and the terrain is punctuated with isolated hills of granites and rhyolites.

The Ghaggar Plain

It consists mainly of former flood plains of river Ghaggar and aeolian deposits. A network of canals covers the entire area The southern and south-eastern part is occupied by medium to high dunes. Nineteen of these interlunar depressions are being utilized for storing the diverted Ghaggar flood waters. The central part of the Ghaggar Plain is drained by the regulated flood waters of Ghaggar River.

Vindhyan Scarpland and Deccan Lava Plateau

The south-eastern plains are locally characterized by plateau, scarp-land and ravines. The Vindhyan scarplands are seen all along the Great Boundary Fault from Chittorgarh to the trijunction of Bharatpur, Dholpur and Sawai Madhopur districts. The Deccan Lava Plateau is mainly confined to parts of Kota, Jhalawar, Banswara and Chittorgarh districts. The ravines, locally impassable, are confined to the alluvium overlying the Vindhyans in Dholpur, Sawai Madhopur, Jhalawar and Kota districts along the Chambal River and its tributaries.

Drainage

The Aravalli Hill Ranges form the main water divide in Rajasthan. Luni is the only river west of Aravallis. In the remaining area of western Rajasthan, comprising about 60 per cent of the geographical area of the State, the drainage is internal, and the streams are lost in the desert sands after flowing for a short distance from the point of origin. Luni itself essentially is an ephemeral stream with flood cycle of 16 years. Drainage in western Rajasthan is towards west and south-west. In the east of Aravalli ranges, the main drainage is towards north-east. The Chambal Catchment occupies 21 per cent (72,032 sq km) of the total geographical area of the state. The other important catchments include Yamuna-Ganga in the north-east, and Mahi and Sabarmati in the south west with flow towards south. The former three catchments support perennial rivers. In the northern and north-eastern parts of eastern Rajasthan, the Banganga, Barah, Sota, Sahibi and Kantli rivers are of inland nature. The drainage in the whole of Rajasthan is generally dendritic. In the desert area, a few salt lakes and depressions exist, prominent among them being the Sambhar Lake, Didwana Lake, Bap, Pachpadra and Rann of Jaisalmer and Pokaran.

Geology

Diverse rock types ranging from the oldest Archaean metamorphic to sub-Recent to recent alluvium and windblown sand are exposed in Rajasthan. Diverse rock types ranging from the oldest Archaeanmetamorphics to sub-recent to recent alluvium and windblown sand are exposed in Rajasthan. However, in a major portion of the area, particularly in western Rajasthan, the older rocks lie concealed below a cover of alluvium and blown sand. The State is underlain by hard rocks (nearly 40%) consisting of the Archaeancrystallines (Bhilwara Super group), Proterozoic rocks comprising Aravalli and Delhi Supergroup, Erinpura Granite, Malanivolcanics and plutonic suite of rocks and their equivalents, Marwars, Vindhyans and Deccan Traps. The soft rocks include the alluvium and the blown sand which occupy the major portion in the remaining part of the State.

Hydrogeology

Hydrogeological map of Rajasthan is presented in Figure 14.2. The principal source of recharge to ground water in Rajasthan is rainfall. For broadly grouping geological formations from ground water occurrence and movement considerations, the various litho units have been classified into two groups on the basis of their degree of consolidation and related parameters:

I. Porous formations
   (a) Unconsolidated formations
   (b) Semi-consolidated formations
II. Fissured formations
   (a) Consolidated sedimentary rocks
   (b) Igneous and metamorphic rocks
   (c) Volcanic rocks
   (d) Carbonate rocks

Porous Formations

The Quaternary sediments comprising of younger as well as older alluvium are the most important unconsolidated formations and form the potential aquifers. These sediments are composed of clay, silt, sand, gravel and mixture of concretions, etc. Sand, gravel and admixture of these form the potential aquifers in northern, eastern, north-eastern, western and south-western parts of the state. Ground water occurs

Figure 14.2

Hydrogeological Map of Rajasthan

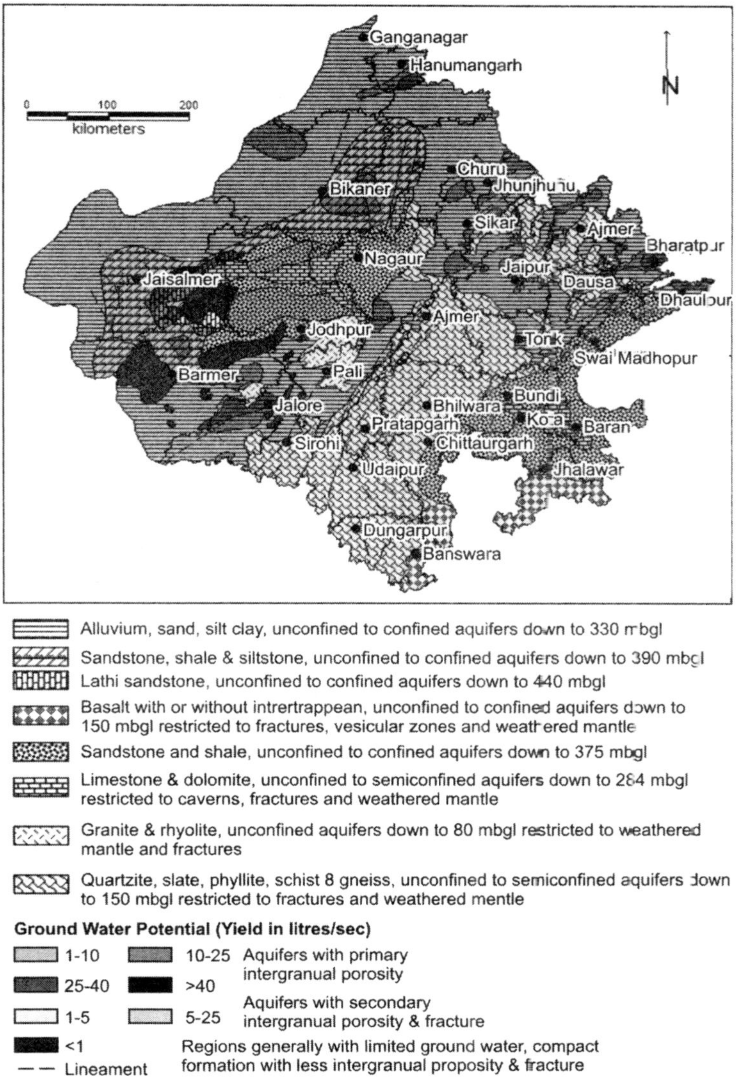

under both water table and confined conditions. The maximum drilled thickness of alluvium is 543.51 metre below ground level (mbgl) at Anupgarh in Ganganagar district.

The semi-consolidated formations belonging to Palaeozoic, Mesozoic and Cainozoic age are composed of siltstone, claystone, sandstone, shale, conglomerate and limestone. Sandstones and limestones form the main aquifers in Jaisalmer, Jodhpur, Barmer and Bikaner districts and encompass Lathi, Jaisalmer, Baisakhi, Bhadesar, Parihar, Abur, Jogisa and Marh formations. Sandstones of Lathi formation are the most potential aquifers in the districts of Jaisalmer, Jodhpur and Barmer. Ground water occurs under semi-confined to confined conditions. Exploratory wells are mostly restricted to 300 m depth and yield of aquifers is generally poor (less than 36 m^3/hr).

Fissured Formations

Fissured formations, as hydrogeological unit, occupy 32 per cent area of the State and can be broadly classified into four units.

Consolidated sedimentary rocks, excluding carbonate rocks, include sandstones and shales. In eastern and south-eastern part of the state these belong to Vindhyan Supergroup, whereas in western Rajasthan these belong to the Marwar Supergroup. Yield of wells is generally poor i.e. below 36 m^3/hr. Some of the Vindhyan sandstone occupying parts of Kota, Jhalawar, Bundi and Chiitorgarh and sandstone of Marwar Supergroup in Jodhpur and Nagaur districts have promising yield ranging from 36 to 90 m^3/hr. These form important sources of ground water supply due to their porous and permeable nature locally.

Igneous and metamorphic rocks of lower Proterozoic age comprise of Slate, Quartzite, Phyllite, Schist, Gneiss and various crystallines of Bhilwara Supergroup have meagre primary porosity. These formations, particularly granites, have developed significant secondary porosity locally due to weathering and fractures. Ground water occurs under water table condition in the weathered mantle and under semi-confined conditions in deep seated fractures. These formations are mostly found in the districts of Banswara, Dungarpur, Udaipur, Chittorgarh, Bhilwara, Tonk, Jaipur, Alwar and Jhunjhunu in eastern Rajasthan and Nagaur, Churu, Barmer, Jaisalmer, Pali, Jalore, Sirohi and Jodhpur districts in western Rajasthan. Exploratory wells drilled are mostly restricted to 200 m depth in general. In most parts of the State, crystallines of Archaean to Upper Proterozoic age have poor yield, i.e. less than 36 m^3/hr.

Volcanic rocks include Deccan Trap Lava Flows and occur in parts of Barmer, Jhalawar, Chittorgarh and Banswara districts. These are basaltic to doleritic in composition. Occurrence and movement of ground water in these formations is controlled by the presence of vesicles, extent of weathering, jointing and fracture pattern. Deccan traps exhibit low to medium permeability depending upon the primary and secondary porosities in them. Exploratory boreholes are restricted to 125 m depth and yield ranges from less than 6 m^3/hr to about 30 m^3/hr. The Coefficient of Transmissivity of aquifers varies from less than 10 to 250 m^2/day.

Carbonate rocks include limestone, marble and dolomite of Proterozoic and Upper Palaeozoic to Mesozoic age and occupy parts of Kota, Bundi, Jaipur, Sawai Madhopur and Alwar districts on the eastern side of Aravallis and parts of Nagaur, Bikaner, Jaisalmer and Jodhpur districts in western Rajasthan. Limestones in some areas have developed solution cavities and yield good quantities of water.

Ground Water Exploration

First large scale exploratory drilling for ground water in Rajasthan was taken up during 1956–57 under the All-India Ground Water Exploration Programme under Indo-US Technical Cooperation Operational Agreement No. 12, wherein 10 exploratory boreholes were drilled in parts of Jaisalmer, Bikaner, Barmer, Churu, Sikar and Jodhpur districts. Subsequently, Geological Survey of India carried out exploratory drilling for water supply to Khetri Copper Project and Saladipura Fertilizer Project in Sikar, Jhunjhunu and Alwar districts between 1961 and 1970. Besides, drilling was also carried out in Banas river basin falling in Udaipur district. During 1967, exploration was carried out in Nagaur district during 1968–69 in Jodhpur and Barmer districts under Drought Scheme.

Since 1967, three UNDP assisted Ground Water Exploration and Assessment Projects of CGWB viz. UNDP Phase I, Phase II and Phase III were implemented in the State. Central Ground Water Board had taken up extensive ground water exploration programmes in different parts of the State in the recent years.

The objective of ground water exploration is to delineate potential aquifer zones and to assess their yield characteristics. A total of 1,530 exploratory, 366 observation wells, 147 slim holes and 530 piezometers have been drilled in the State till March, 2013, including 843 exploratory wells in unconsolidated and 687 in consolidated formation. Status of ground water exploration as on 31 March 2013 in the State is furnished in Table 14.1.

Table 14.1

District-wise Status of Exploratory Wells Drilled in the State (as on 31.3.2013)

District	*Geographical area (sq.km)*	*Ground water exploration*				*Formations*	
		EW	*OW*	*SH*	*PZ*	*Unconsolidated/semi-consolidated*	*Consolidated*
Ajmer	8,481	47	3	1	6	5	42
Alwar	8,380	55	7	5	43	36	19
Banswara	5,037	43	4	0	0	0	43
Baran	6,992	7	2	0	0	1	6
Barmer	28,387	68	8	5	22	64	4
Bharatpur	5,066	57	26	6	38	37	20
Bhilwara	10,455	39	4	1	0	9	30
Bikaner	27,244	43	9	2	62	43	0
Bundi	5,776	22	3	5	0	3	19
Chittorgarh	10,856	43	10	0	2	0	43
Churu	16,830	51	16	12	28	51	0
Dausa	3,432	41	7	3	29	29	12
Dholpur	3,033	5	5	4	11	4	1
Dungarpur	3,770	28	8	0	0	0	28
Ganganagar	10,978	19	36	4	18	19	0
Hanumangarh	9,656	43	9	3	20	43	0
Jaipur	11,143	100	36	1	55	81	19
Jaisalmer	38,401	109	24	17	16	109	0
Jalore	10,640	26	17	21	20	18	8
Jhalawar	6,219	25	3	3	1	0	25
Jhunjhunu	5,928	95	13	0	39	84	11
Jodhpur	22,850	71	16	13	34	49	22
Karauli	5,524	33	0	3	12	10	23
Kota	5,217	25	2	1	1	2	23
Nagaur	17,718	54	16	22	10	35	19
Pali	12,387	83	6	0	1	8	75
Rajsamand	3,860	43	2	0	4	8	35
Sawai Madhopur	4,498	48	22	7	3	12	36
Sikar	7,732	97	34	2	45	83	14
Sirohi	5,136	22	3	0	1	0	22
Tonk	7,194	21	3	6	7	0	21
Udaipur	13,419	67	12	0	2	0	67
Total	3,42,239	1,530	366	147	530	843	687

Source: Ground Water Department.

Exploratory drilling in Rajasthan has brought out significant results in terms of aquifer characteristics, ground water potentiality and quality scenario in shallow and deeper aquifers. Alluvium contains the most productive aquifers but in extensive areas ground water quality is poor. Highly potential alluvial aquifers are met with locally in Sikar, Jhunjhunu, Churu, Jalore, Barmer, Bharatpur, Bikaner, Alwar and Nagaur districts having yield from 144 m^3/hr to 360 m^3/hr. Coefficient of Transmissivity of the aquifers at places ranges up to 2,000 m^2/day and more, and the Specific Capacities of the tube wells range upto 750 m^3/hr/m of drawdown in these districts. Ground water exploration in Bharatpur and Jaisalmer districts has enabled demarcation of an artesian flowing belt. However, at present flowing conditions do not exist. Highly potential alluvial valley fills are also met within the hard rock terrains falling in Jhunjhunu, Bhilwara, Udaipur and Tonk districts, particularly along and within the river channels, although saturated thickness is generally limited in these valley fills. The transmissivity of the aquifer in the valley fills along Kantli River in Jhunjhunu district ranges from 2,000 to 7,000 m^2/day and the Specific Capacity of wells range from 9 to 90 m^3/hr/m of drawdown. Valley fills in Tonk, Bhilwara and Udaipur districts exhibit very high Transmissivity ranging from 2,500 to 5,000 m^3/day.

Presences of moderate to highly potential aquifers in the semi-consolidated formations are in existence. In Jaisalmer, Barmer, Bikaner and Churu districts are having discharge up to 360 m^3/hr. Exploration under the Indo-US, TCM Project in Rajasthan revealed, for the first time, high ground water potential of Lathi Series of Jurassic age falling in Jaisalmer and Barmer districts. Lathi aquifer is capable of yielding 186–209 m^3/hr for drawdown of 2–10 m. Transmissivity ranges between 996 and 1,634 m^2/day. Storage Coefficient ranges between 9.6×10^{-4} and 1.3×10^{-4}. Exploration carried out under UNDP Project phase-I in Borunda area of Jodhpur district indicated that Bilara Limestone of Marwar Supergroup is highly certified with development of caverns. These limestones have a maximum thickness of 100m. Fresh ground water occurs over an area of 161 sq km and the aquifer is unconfined with water level depth ranging from 4 to 32 mbgl. Nagaur Sandstone and Jodhpur Sandstone of Marwar Supergroup are found to be moderately potential. Wells tapping Jodhpur Sandstone down to depth of 60 to 107 m yield 30 to 70 m^3/hr of ground water.

Ground water exploration in consolidated formations reveals low potentiality of aquifers in general. However, limestone and sandstone of Vindhyan Supergroup and Deccan Trap basalts in Kota, Baran,

Bundi, Nagaur, Jodhpur and Jaisalmer districts and Jhalawar and Banswara districts respectively exhibit moderate potentiality with yield ranging from 20 to 36 m^3/hr at drawdown less than 10 m.

Analytical data of exploratory drilling have indicated that the chemical quality of shallow ground water varies from fresh in the southern and eastern parts to highly saline in the western and central parts of the state. Electrical conductivity of ground water ranges from 250 to over 10,000 micro mhos/cm at 25 °C in the western and central parts. Ground water quality in major parts of western Rajasthan is generally saline. Saline waters are mostly of sodium chloride (NaCl) type. Shallow ground water is generally more mineralized in western Rajasthan. In a major part of area in western Rajasthan, shallow ground water has electrical conductivity exceeding 5,000 mmhos/cm at 25 °C. Such saline tracts also occur in Bharatpur, Ajmer, Jhunjhunu, Jaipur and Churu in eastern and Nagaur in central parts of the state. Locally, in western Rajasthan, ground water is found to be highly mineralized ranging up to 80,290 micro mhos/cm at 25 °C. High salinity in ground water is generally confined to aquifers in alluvium, blown sand and semi-consolidated formations encompassing Tertiary and Mesozoic formations, and to a lesser extent in the aquifers of Marwar Supergroup of rocks.

Salinity of ground water in deeper aquifers is highly variable. Quality of ground water varies widely with depth in alluvium and semi-consolidated formations, including Mesozoics and Tertiaries. In general, ground water quality deteriorates with depth but locally the quality improves vertically, and in a few cases depth wise quality of water is highly variable.

Western desertic terrain predominantly is underlain by saline ground water at all levels except local fresh water pockets (Ranau to Kishangarh and area around Ghotaru in Jaisalmer district). The belt extends adjacent to international border with Pakistan from south of Ganganagar to north of Jaisalmer district also covering parts of Bikaner, Churu, Hanumangarh, Jhunjhunu, districts. Major parts of Barmer, Pali and isolated pockets of Jodhpur, Jalore, Nagaur and Bharatpur districts also depict such situations.

Occurrence of saline ground water below fresh ground water is a general phenomenon in the entire State. Western Rajasthan contains multi-aquifer system with vertical variation in degree of salinity. At most places fresh ground water overlies the saline water with saline –fresh water interface at less than 20 m to about 50 m depth. Such conditions are encountered in parts of the Ganganagar, Hanumangarh, Bikaner, Nagaur, Jalore, Churu, Jalore, Bharatpur and in localized pockets of Sawai Madhopur, Karauli and Alwar districts. Fresh water saline water interface at depth range of 50 to 200 m occurs south of Bikaner and Jaisalmer; south-west of Barmer and in western part of the Jalore district.

Occurrence of saline water over fresh ground water has been observed in northern parts of Jodhpur, Nagaur, Alwar districts; north east of Jaisalmer, north of Barmer and south and east of Bharatpur. However, potable water is found in areas covered by sandstone and limestone of Marwar Supergroup, Lathi formation in Barmer and Jaisalmer districts and Tertiaries in parts of Bikaner, Nagaur, Churu, Barmer and Jaisalmer districts, and localized pockets in Quaternaries. Thickness of fresh water zone is less than 10 m towards the eastern margin of the desert resulting in ground water drought situation. Thick fresh water aquifers exist in Jaisalmer district (Lathi basin and Palaeo-courses of lost Saraswati (around Tanot-Kishangarh and Ghotaru areas), Bikaner and Nagaur district (Palana-Nagaur basin), Barmer district (Bhadka area), Jodhpur district (Borunda limestone) etc.

Areas with fluoride exceeding 1.5 mg/l occur in almost all the districts of Rajasthan and under diverse hydrogeological conditions. High concentration of fluoride in the ground water is widespread in western Rajasthan; inter alia appears to be related to aridity. Fluoride exceeding 1.5 mg/l has also been noticed in ground water in semi-arid tracts of eastern Rajasthan.

Ground Water System Behaviour

In order to study the behaviour of ground water regime, national network of hydrograph stations (NHS) have been established in the State of Rajasthan. These wells are monitored four times in a year during the months of January, May, August and November. Water samples are collected for quality monitoring once in a year during the month of May. *Ground Water Year Book*, incorporating the data generated under the ground water monitoring is brought out periodically on annual basis.

Depth to Water Level During Pre-monsoon

The depth to water level varies widely throughout the State. Depth to water level map during pre-monsoon (May 2014) is presented in Figure 14.3. Overexploitation and excessive withdrawal of ground

Figure 14.3

Depth to Water Level Map of Rajasthan State (May, 2014)

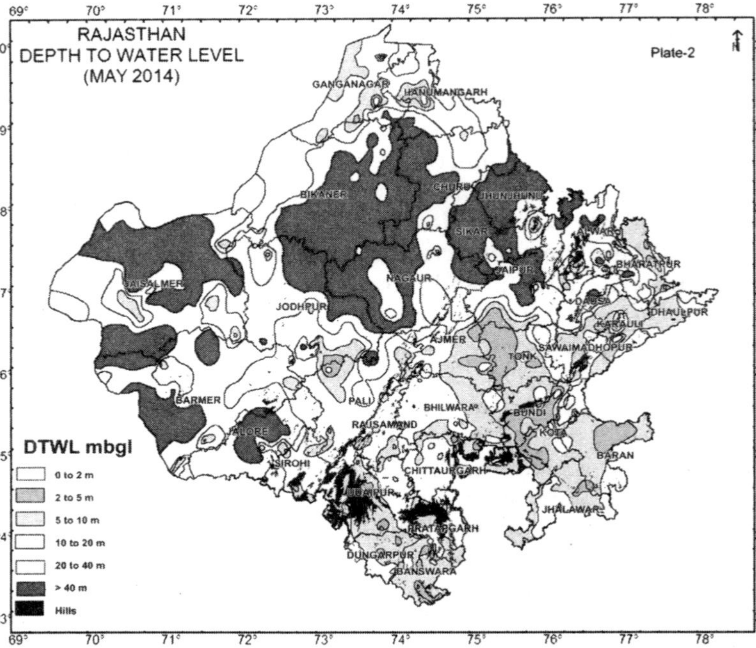

Source: Central Ground Water Board, 2015.

water have led to substantial decline in water levels, which may ultimately result in drying up of aquifers in many areas of the State.

Fluctuation between May 2013 and May 2014

Comparison of water level data of May 2014 with that of May 2013 (Figure 14.4) indicates that there has been decline in water level in the range of 0 to 2 m during May 2013 in major part of the state.

Long-Term Water Level Fluctuation

Water level data of May 2014 when compared with Decadal Mean of May (2004–2013) (Figure 14.5) indicates rise in water levels in major part of the State except central belt running from Jaisalmer in the west to Bharatpur in the east and a few localized pockets in Pratapgarh,

Figure 14.4

Water Level Fluctuation during Pre-monsoon (May, 2013 – May, 2014)

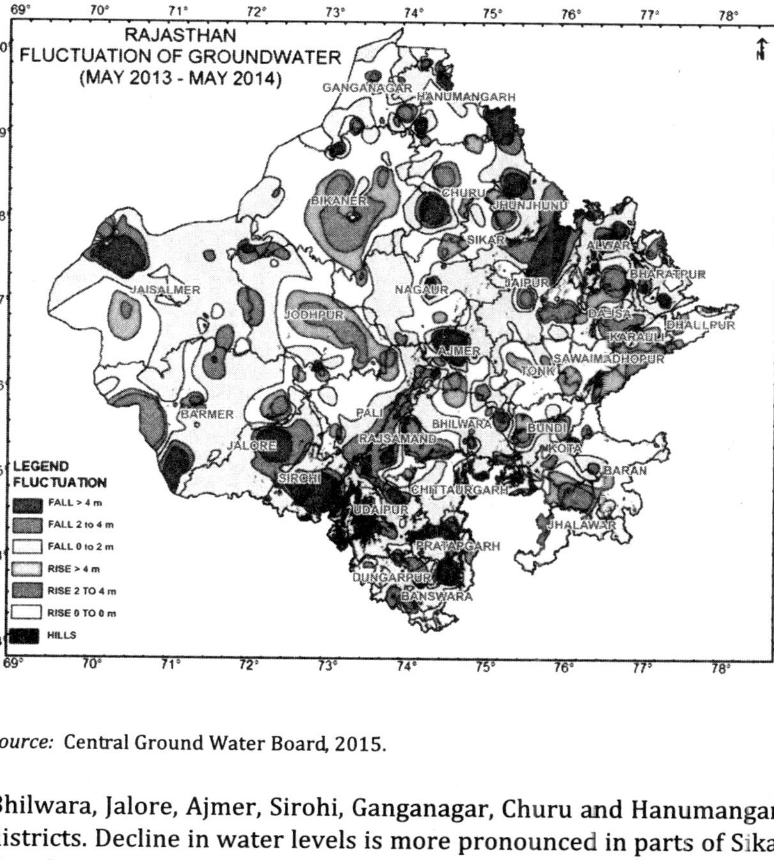

Source: Central Ground Water Board, 2015.

Bhilwara, Jalore, Ajmer, Sirohi, Ganganagar, Churu and Hanumangarh districts. Decline in water levels is more pronounced in parts of Sikar, Jhunjhunu, Jaipur, Alwar, Bikaner, Nagaur and Jodhpur districts.

Ground Water Resources and Development Potential

District-wise ground water resources, jointly estimated by the Central Ground Water Board and State Ground Water Department as on 2011 are presented in Table 14.2. Estimates reveal the fact that scope for future ground water resource development in the state of Rajasthan is very less. Taranagar block in Churu district has not been assessed due

Figure 14.5

Decadal Water Level Fluctuation Map of Rajasthan

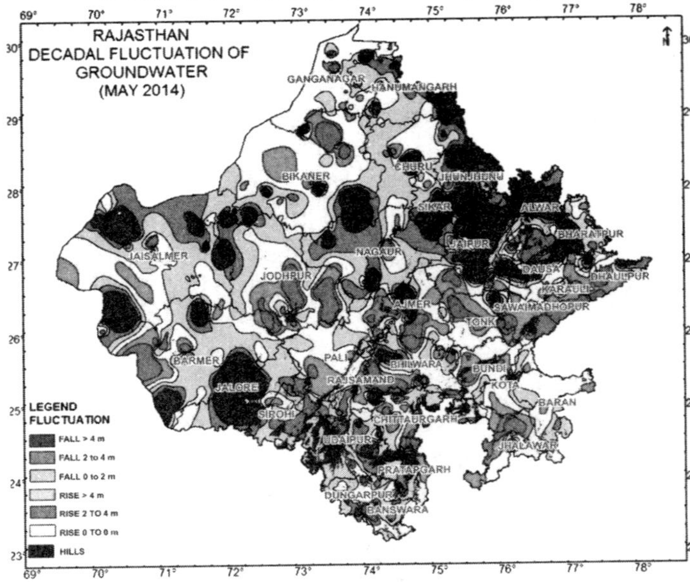

Source: Central Ground Water Board, 2015

Figure 14.6

Categorization of Assessment Units (Blocks) as on 31 March 2011

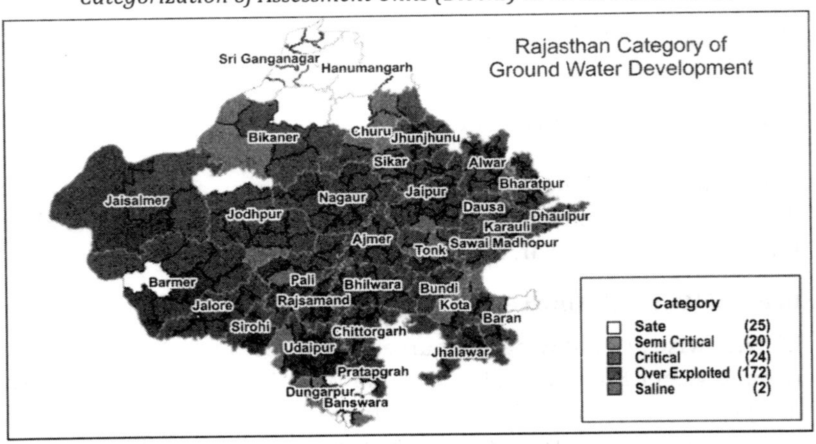

Source: Central Ground Water Board, 2015

Explanation about Semi-critical, Critical and Over-exploited areas as per guidelines of Ground Water Estimation Committee is as follows:

Table 14.2

Ground Water Resources of Rajasthan State as on 31 March 2011

Assessment Unit/District	Area	Potential Area	Total Annual Replenishable Water Recharge	Natural Discharge during Non-Monsoon Season	Net Annual Ground Water Availability	Existing Gross Ground Water Draft for Irrigation	Existing Gross Ground Water Draft for Domestic and Industrial Water Supply	Existing Gross Ground Water Draft for all Uses	Net Ground Water Availability for future Irrigation Development	Ground Water Allocation for Domestic and Industrial Development	Stage of G.W. Development	Category
	Sq. km	Sq. km	MCM	MCM	MCM	MCM	MCM	MCM	MCM	MCM	%	
Ajmer	8,481.00	7,466.76	355.11	33.05	322.06	415.47	46.91	462.38	00.00	46.91	143.57	OE
Alwar	8,720.46	6,825.21	795.51	63.82	731.70	1215.52	95.66	1311.18	3.99	95.95	179.20	OE
Banswara	4,536.08	3,979.96	257.62	26.96	230.66	92.31	19.61	111.92	166.87	28.53	48.52	Safe
Baran	6,955.31	6,892.21	519.67	48.04	471.63	525.96	39.38	565.34	55.62	44.75	119.87	OE
Barmer	28,387.00	12,734.65	272.53	25.24	247.29	239.01	67.66	306.66	21.17	68.47	124.01	OE
Bharatpur	5,044.10	3,412.52	492.50	43.14	449.36	458.39	63.78	522.18	3.96	65.69	116.21	OE
Bhilwara	10,455.00	9,354.85	464.32	44.76	419.55	501.98	37.96	541.78	0.00	39.80	129.13	OE
Bikaner	30,381.75	13,602.51	252.65	12.63	240.01	258.67	83.74	342.41	40.11	90.33	142.66	OE
Bundi	5,500.00	4,240.18	422.65	59.70	362.95	336.69	27.11	363.80	53.28	34.72	100.23	OE
Chittorgarh	7,880.06	6,095.00	352.76	34.17	318.60	432.27	13.79	446.06	00.00	13.79	140.01	OE
Churu	13,792.95	5,191.74	141.01	7.05	133.95	94.45	24.08	118.54	30.96	43.98	88.49	S.Critical
Dausa	3,420.17	3,085.62	263.73	26.20	237.53	378.86	24.87	403.74	0.00	24.87	169.97	OE
Dholpur	3,009.05	2,486.14	283.12	23.82	259.31	307.06	25.26	332.33	15.96	25.76	128.16	OE
Dungarpur	3,770.00	2,634.13	141.03	12.91	128.12	83.96	8.82	92.78	0.00	44.16	72.41	S.Critical

Cont'd...

...Cont'd

Ganganagar	11,603.65	1,545.64	407.98	40.80	367.19	156.17	5.45	161.63	200.68	10.33	44.02	Safe
Hanumangarh	9,579.54	1,278.50	226.49	22.65	203.84	157.16	7.05	164.21	34.46	11.07	80.56	Safe
Jaipur	11,061.44	9,994.67	712.38	66.85	645.53	1081.11	256.86	1337.97	9.79	257.33	207.27	OE
Jaisalmer	38,401.00	12,090.04	67.92	6.33	61.59	93.56	28.93	122.49	18.74	29.61	198.89	OE
Jalore	10,640.00	8,228.10	462.85	40.86	422.00	779.52	40.90	820.42	9.69	41.91	194.41	OE
Jhalawar	6,219.00	6,106.16	441.71	29.48	412.23	476.01	15.83	491.83	4.79	19.40	119.31	OE
Jhunjhunu	5,928.00	5,273.69	263.51	23.50	240.01	456.04	85.71	541.75	3.30	86.04	225.72	OE
Jodhpur	22,250.00	18,867.92	429.52	42.39	387.13	727.00	103.97	783.33	47.61	113.03	202.34	OE
Karauli	5,038.60	3,902.42	373.14	36.17	336.97	411.77	50.64	462.41	12.39	49.84	137.23	OE
Kota	5,203.94	5,123.17	566.89	53.59	513.30	420.97	48.91	469.88	77.76	62.51	91.54	Critical
Nagaur	17,718.25	16,378.50	580.75	56.49	524.26	811.86	178.93	990.79	41.20	189.53	188.99	OE
Pali	12,357.00	7,362.54	328.59	32.29	296.31	314.36	27.47	341.84	9.59	30.48	115.37	OE
Pratapgarh	4,359.80	2,950.39	155.21	14.36	140.86	168.84	5.97	174.81	3.99	9.82	124.10	OE
Rajsamand	4,635.46	3,540.09	118.72	11.87	106.85	102.70	15.25	117.95	0.63	15.25	110.39	OE
Sawai Madhopur	5,020.65	7,263.46	321.48	30.37	291.11	370.52	58.42	428.94	2.09	83.75	147.35	OE
Sikar	7,880.85	4,325.63	396.36	35.94	360.42	374.08	79.47	453.55	9.72	62.38	125.84	OE
Sirohi	5,136.00	4,075.70	303.38	28.54	274.85	299.52	11.08	310.61	11.26	14.84	113.01	OE
Tonk	7,200.00	6,524.72	483.53	44.59	438.95	357.45	76.35	433.80	18.41	91.04	98.83	Critical
Udaipur	11,760.60	7,770.92	286.85	33.94	252.90	229.31	31.95	261.25	4.50	39.46	103.30	OE
State Total	3,42,326.71	2,20,603.74	11,941.45	1,112.49	10,828.97	13,133.1818	1,709.8153	14,842.9972	912.50	1,885.32	137.07	OE

Source: Centre Ground Water Department.

to poor quality in the entire block. District-wise number of blocks under each category is depicted in Figure 14.6.

Semi-critical Areas

Areas where ground water resource assessment shows the stage of ground water development more than 70 per cent but less than 90 per cent and either pre-monsoon or post-monsoon ground water level shows a significant long term decline.

Areas where ground water development stage is more than 90 per cent but neither pre- nor post-monsoon ground water level shows significant long-term decline.

Areas where stage of ground water development is less than 70 per cent and both pre-monsoon and post-monsoon water levels show significant long-term decline.

Critical Areas

Areas where stage of ground water development is more than 90 per cent and either pre- or post-monsoon water level shows significant long-term decline.

Areas where stage of ground water development is more than 70 per cent and less than 100 per cent and both pre-monsoon and post-monsoon water levels show significant long-term decline.

Overexploited Areas

Areas where stage of ground water development is more than 100 per cent and both pre- and post-monsoon water levels show significant long-term declining trend.

Rajasthan State witnesses wide spectrum of hydrogeological conditions as well as availability of ground water resources. Ground water recharge in Thar desert areas of Western Rajasthan is mostly less owing to arid climatic conditions (low rainfall and high evaporation) and ground water development is also relatively less due to constraints of deep ground water levels and inferior quality of ground water as well as availability of canal water in parts. Due to availability of canal water and poor quality of ground water, Ganganagar and the Hanumangarh districts fall in safe category. Hard rock areas of Aravalli hills are vulnerable to water crisis during spells of drought due to their limited scope for dynamic recharge to ground water. Ground water development in alluvial plain areas especially on either side of Aravalli

Hill Range is on higher side and most of the blocks underlain by alluvium fall in over-exploited category.

Ground Water Development Over the Years

Figures 14.7, 14.8 and 14.9 depicts ground water development, irrigation draft over the years and categorization of blocks with time according to ground water development respectively.

Figure 14.7

Stage of Ground Water Development from 1984 to 2011 Assessment

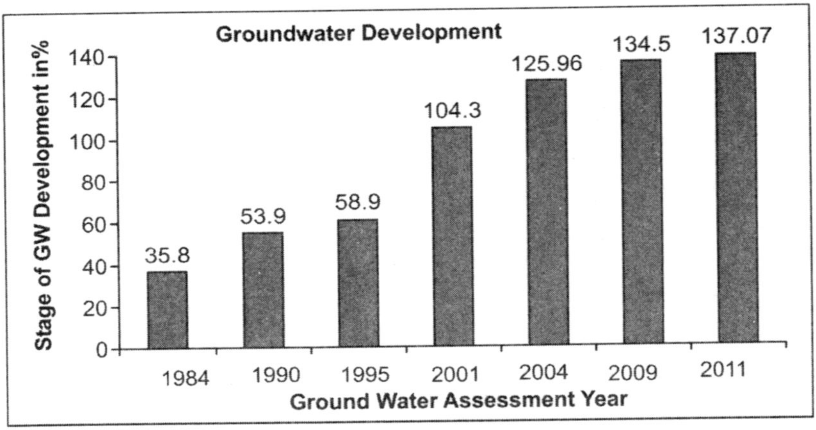

Figure 14.8

Irrigation Ground Water Draft from 1984 to 2011 Assessment

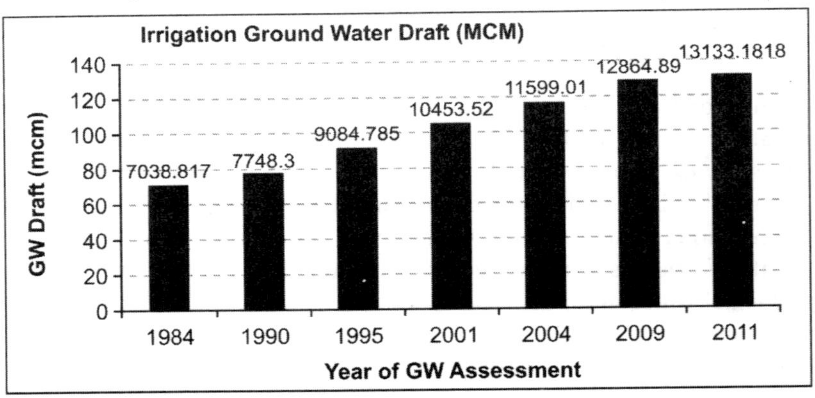

Figure 14.9

Categorization of Blocks with Time

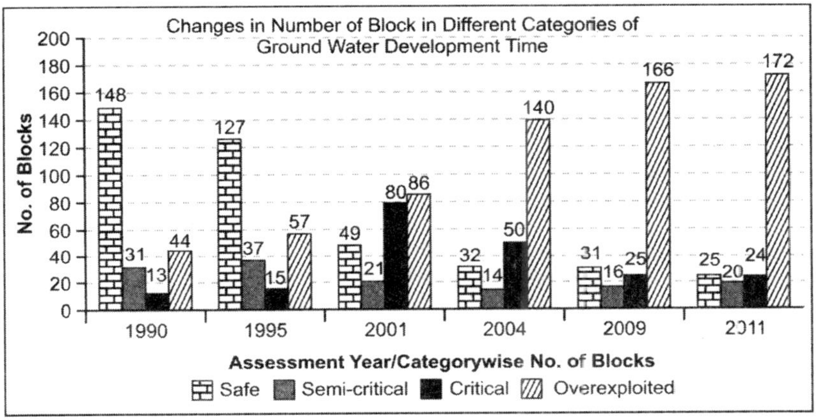

Ground Water Quality

Figure 14.10, 14.11 and 14.12 shows the Electrical Conductivity, Fluoride (F) and Nitrate (NO_3) respective values and distribution in the state.

Figure 14.10

ISO Electrical Conductivity Map of Rajasthan

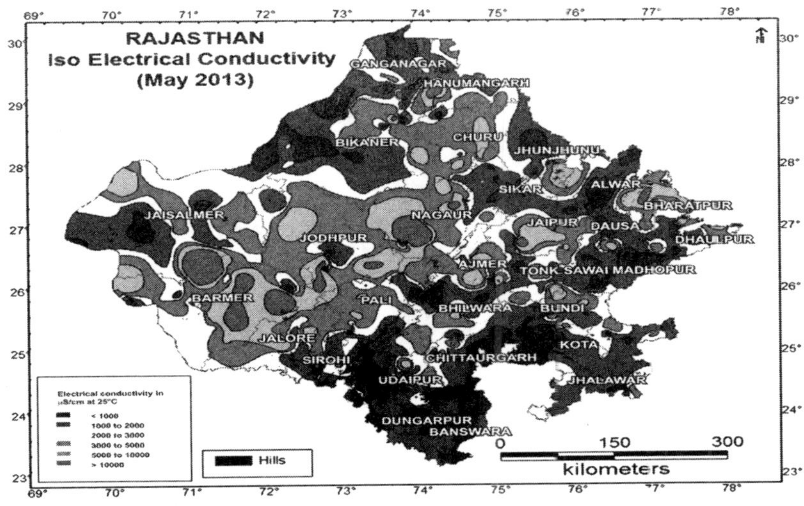

Source: Central Ground Water Board, 2015

Figure 14.11

Fluoride Distribution Map of Rajasthan

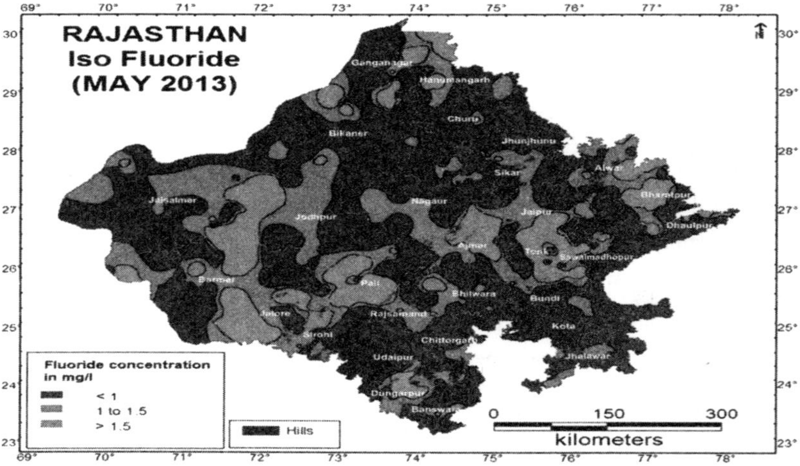

Source: Central Ground Water Board, 2015

Figure 14.12

Nitrate Distribution in Rajasthan

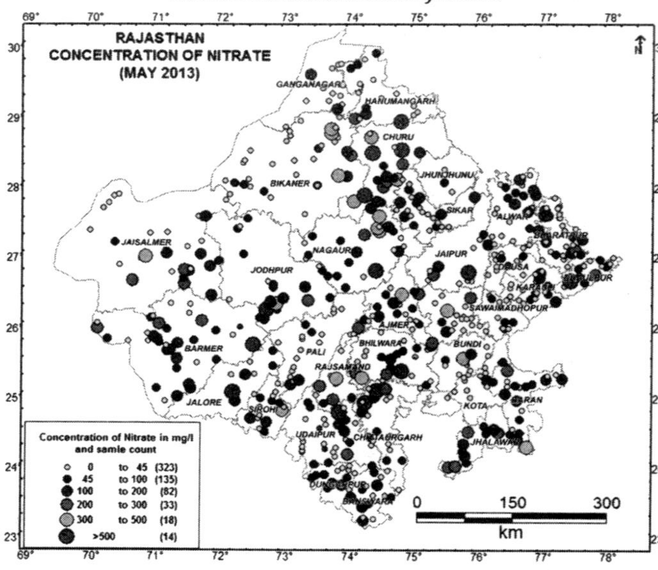

Source: Central Ground Water Board, 2015

Ground Water Related Issues and Problems

A substantial part of the State is covered by hard rock formations which have poor water yielding capacity. The State, especially western Rajasthan, has very deep water levels, therefore agricultural activities in this area are by and large restricted to IGNP canal command areas only or flood plains of Luni catchment where shallow water levels are available. Rapid declining trends in ground water levels are due to over-exploitation in other areas. Waterlogging and soil salinity/alkalinity problems have occurred due to seepages from canals, return seepage from irrigation in Indira Gandhi Nahar Pariyojana as well as Chambal canal command areas. Ground water resources in larger Rajasthan are saline. Aridity, diversified geo-formations, inland drainage, waterlogging situations etc. are the root causes for salinity in ground water and development of salt lakes in the state. High fluoride content in ground water is one of the main constraints in almost all the districts of the State. Lack of proper sewerage systems, solid waste disposal in urban areas and excess use of fertilizers for intensive cropping are the main causes of deterioration in chemical quality of ground water by enhancing nitrate concentration etc. is due to lack of effective disposal system.

Conclusion

Rajasthan State witnesses wide spectrum of hydrogeological conditions as well as availability of ground water resources. Ground water resources in larger Rajasthan are saline and it also has poor water yielding capacity. Rapid declining trends are evident in ground water levels due to overexploitation. Waterlogging and soil salinity/ alkalinity problems, lack of proper sewerage systems, solid waste disposal in urban areas and excessive use of fertilizers for intensive cropping are the main causes of deterioration in chemical quality of ground water. Lack of effective disposal system has resulted in enhancement of nitrate concentration.

Irrigation draft which was 7,038.817 mcm in 1984 assessment year has now increased to 13,133.1818 mcm according to a 2011 assessment. Stage of ground water development of Rajasthan was 35.8 per cent during 1984 assessment and it has increase to 137.07 per cent according to latest 2011 assessment. The stage of ground water development has increased mainly due to overdraft. During 1990 assessment, 148 blocks were falling in safe category, which are now reduced to 25 blocks. The number of overexploited blocks has increased from 44 to 172 as per 2011 assessment.

As per analytical results of ground water samples collected during May, 2013, Electrical Conductivity values have been found to range from 200 to 75,000 ìS/cm, this conforms broadly with chloride concentration.

Fluoride contamination in ground water is endemic in the state. High fluoride content in ground water is one of the main constraints in almost all the districts of the State. It is the reason of increasing number of fluorosis patients. Nitrate concentration is also very high in the state, further increasing poor quality of water and negative effect on human health.

To combat the endemic and anthropogenic problems, actions must be taken to address the problem, like conjunctive use of ground water and canal water to avoid land degradation by waterlogging and soil salinity/alkalinity, use of desalination/de-fluoridation technology (Phytoremediation and Bioremediation), enforcement of ground water regulation and implementation of suitable management strategies, low water intensive cropping pattern and adoption of sprinkler, drip irrigation techniques, close field distribution channels etc. should be promoted, artificial recharge in depleted aquifers needs to be promoted, construction of small check dams or earthen dams, at suitable sites, to store rainwater, watershed development and soil conservation practices should be encouraged. Deepening of dug wells in hard rock formation should be carried out to have enough storage during pumping and to avoid drying up of such wells during summers; horizontal drilling can be done to tap lateral fractures for enhancing the yield of wells, industrial effluents should be treated before discharging into the natural water bodies/rivers etc. and most importantly, awareness programmes about conservation of precious ground water resources and training on rainwater harvesting should be organized regularly throughout the state.

15

Understanding Linkage of Climate-Water-Health in Banas Basin of Rajasthan

Devesh Sharma and Swatantra Kumar Dubey

According to the IPCC (4th Assessment Report), the climate change is very likely due to anthropogenic activities, increase in greenhouse gases, increase in temperature and other meteorological variables (IPCC, 2007). In the last few decades, the climate change issues is globally accepted which is quite visible in the form of undesirable changes in environmental conditions and societal issues (Modiri and Kamali, 2012). In the arid and semi-arid region, the climate conditions are more variable and uncertain which results in frequent occurrence of drought, damage to the environment, water resources scarcity, degradation in soil and vegetation situation in the region (Modiri and Modiri, 2016). In various parts of the world, most of the scientist also analysed the historical and future trends in temperature and precipitation variables (Sharma and Babel, 2014; Ahmad et al., 2015). It is important to understand the existing situation of climate change at various levels like global, regional, national and basin level. According to World Health Organization (WHO), there is a decreasing trend in the precipitation and an increasing trend in temperature (WMO, 1996). The average temperature of the earth increased approximately 0.85 °C during the period 1880–2012 and temperature will further increase between 0.3 °C and 4.8 °C by the end of twenty-first century (Hartmann et al., 2013; IPCC, 2014). The historical climatic data are useful in assessing the environmental risk, including the drought, flood, soil erosion and other hazards (Parajka et al., 2010). Some studies demonstrated that mean temperature was increasing over the last century at different regions of the world at different spatial scale (Vose et al., 2005; Toreti and Desiato, 2008; Klok and Tank, 2009; Del Rio et al., 2012; Pellicone et al., 2014; Caloiero, 2015). Alexander et al. (2006) analysed the global annual trends in extreme rainfall and found an increasing trend in the precipitation indices and the decline trend in the annual number of consecutive dry days. In the past, several studies were carried out on different spatial and temporal scales over India for studying trends and variability of temperature and precipitation, extreme indices (Kothawale et al., 2010; Panda et al., 2014). Sen Roy and Balling (2004) explained the most of the time series shows the increasing trends in extreme precipitation indices in the Indian context. Kumar et al. (2011) used the high-resolution regional climate model (PRECIS) over India to indicate the summer monsoon climate. The results showed that the summer monsoon precipitation will increase by 9–16 per cent in 2080s as compared to the baseline period 1961–1990 under the A1B scenario. Ray et al. (2009) studied climate variability and extreme weather events for Gujarat state using the two indices i.e. cold wave and heat wave.

Recently, there are also concern among scientist to understand the possible linkage between the climate change and health. Young et al. (2004) reviewed the direct and indirect disaster associated with environmental contamination and affect the human health and examined the historical disaster for improving the mitigation and preventive measure. Akhtar (2007) investigate the effect of heat wave in relation to human mortality pattern in India and gives the relation between rainfall and malarial diseases at city level. Kovats and Akhtar (2008) reviewed the impacts of temperature, rainfall and extreme events on health, particularly the impacts of heat waves and floods, and described the risks to health from extreme weather events and improving the effectiveness of public health interventions. Climate change resilience required the improvement of infrastructure. According to Health Dimension of Climate Change report, the extreme events like heatwave, flood, and drought have their huge impact on health of human and society as given in Table 15.1.

Table 15.1

Extreme Weather Events and Related Health Impacts

Climatic Events and Diseases			
	Heatwave	*Flood*	*Drought*
Mortality (Cause-specific)			
Drowning		*	
Physical Trauma		*	
Heat Exhaustion	*	*	
Fire	*		*
Morbidity (System-Related)			
Respiratory Diseases			
Acute Lower Respiratory Tract Infections			*
Mental Diseases			
Depression		*	*
Post-Traumatic Stress		*	*
Morbidity (Transmission Pathway-Related)			
Vectorborne Diseases			
Leptospirosis		*	
Malaria		*	
Waterborne Diseases			
Cholera		*	
Dysentery		*	
Hepatitis A		*	
Salmonella		*	
Other Diseases			
Malnutrition			*
Dehydration	*		*

Source: Health Dimension of Climate Change (World Bank).

Study area

The Banas river basin is located in east-central Rajasthan, between latitudes 24°15' and 27°20' N and longitudes 73°25'E and 77°00'E as shown in Figure 15.1. It originates in the Khamnor hills of the Aravali range and flows along its entire length through Rajasthan (Department of Water Resources). It is bounded by the Luni Basin in the west, the Shekhawati, Banganga and Gambhir basins in the north, the Chambal basin in the east, and the Mahi and Sabarmati basins in the south. The total catchment area is about 45,833 km^2 with a length of about 512 km (Department of Water Resources). The Banas River passes through the

13 districts namely, Sawai Madhopur, Jaipur, Ajmer, Tonk, Rajsamand, Banswara, Chittaurgarh, Udaipur, Bhilwara, Dausa, Nagaur, Sikar and Bundi. The mean annual rainfall over the Banas River Basin is around 541 mm.

Figure 15.1

Location of Study Area and Meteorological Stations Banas River Basin

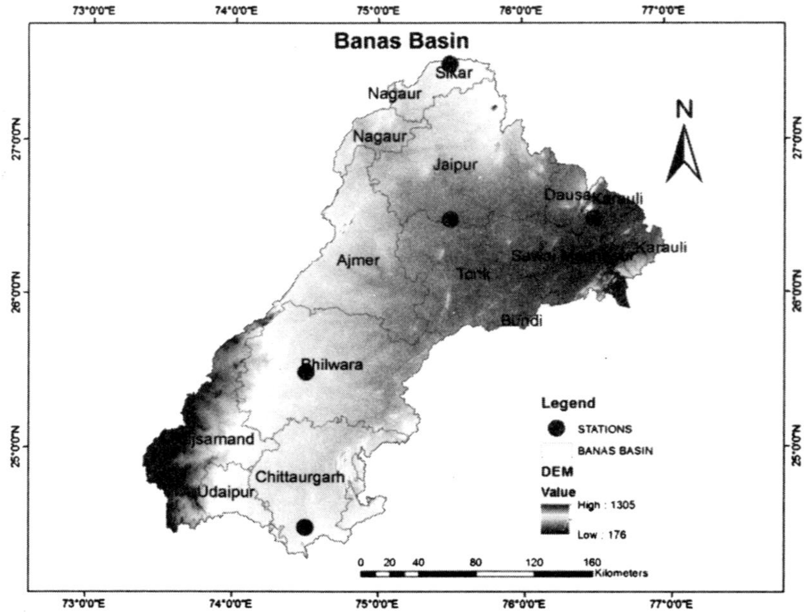

Methodology

The methodology of the present paper is divided into two sections, i.e. climate indices analysis and developing the linkage between climate indices and possible impacts on water resources and human health. For the study, temperature and precipitation data of 5 meteorological stations located in the Banas River Basin are used. The locations of the 5 stations are shown in Figure 15.1 and detail is given in Table 15.2. Two stations are located in the upper part of the basin (Chittorgarh and Bhilwara) and remaining three stations are located in the lower basin. The data were collected from the Indian Meteorological Department (IMD) with spatial resolution of 1x1 degree and period of data is 1972–2013.

Table 15.2

Location Detail of Meteorological Stations

Station Code	*Latitude (North)*	*Longitude (East)*	*Elevation (mm)*	*Location in the Basin*	*Representing Districts*
S1	24.5	74.5	462	Upper Basin	Chittaurgarh
S2	25.5	74.5	461	Upper Basin	Bhilwara
S3	27.5	75.5	488	Upper Basin	Sikar/Nagaur
S4	26.5	75.5	311	Lower Basin	Jaipur/Tonk
S5	26.5	76.5	258	Lower Basin	Sawai Madhopur/ Karauli/Dausa

The RclimDex software package, developed at the Climate Research Branch of the Meteorological Service of Canada (http://etccdi.pacificclimate.org/software.shtml) was used for developing the precipitation and temperature indices using the daily data. In this study, six indices are used as listed in Table 15.3. Out of six indices, four represent precipitation conditions and two represent temperature conditions in the study area.

Table 15.3

Definitions of the Extreme Precipitation and Temperature Indices

Indices	*Units*	*Definitions*
Precipitation Indices		
CDD	Day	Maximum number of consecutive days with RR<1mm
CWD	Day	Maximum number of consecutive days with RR>=1mm
PRCPTOT	mm	Annual total PRCP in wet days (RR>=1mm)
RX1day	mm	Monthly maximum one day precipitation
Temperature Indices		
TN10p	Day	Percentage of days when TN<10th percentile (cool nights)
TX90p	Day	Percentage of days when TX>90th percentile (warm days)

Note: TN represents minimum temperature and TX represents maximum temperature.

Figure 15.2 shows the linkage framework to understand the impact of climate change on water resources and further effects on health and society. Similar approach is used in the paper for the selected districts of Banas Basin.

Figure 15.2

Flow Chart of Climate, Water and Health Impact

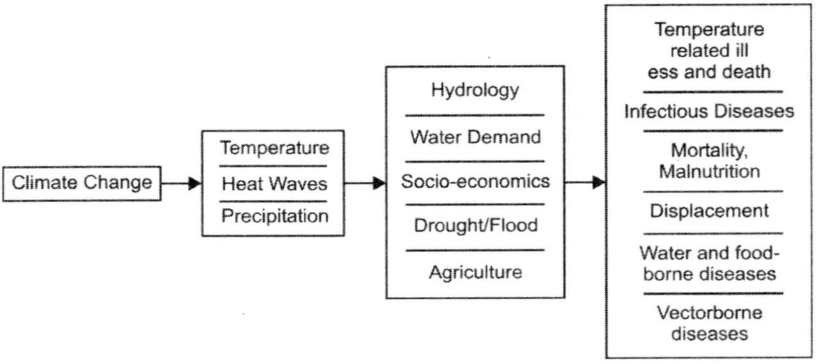

Source: Adapted from *Health Dimension of Climate Change Report*, World Bank.

Results and Discussion

Trend analysis was done for both temperature and precipitation for all the five meteorological stations. All the six indices (four indices for precipitation and two indices for temperature) are analysed for the period 1972–2013. Figures 15.3 to 15.8 show the temporal variation in all the six indices for five stations.

There is an increase trend is observed in the CDD index for all the station as shown in Figure 15.3. It means that there is an increasing trend in the number of wet days and dry days in the basin as whole. Figure 15.4 shows the time series and variation in the CWD index with increased trend at all the five stations. There is a sudden positive change after year 2010 as shown in figure. Similar On the other side, total annual precipitation is decreasing at four stations, but increasing at Sikar station (Figure 15.5). It interprets that there may be more chance of more rainy days with little rain, but overall decrease in amount for four stations. There is a decrease in monthly maximum 1-day precipitation for all the stations which inference that amount of maximum daily rainfall is decreasing (Figure 15.6). Figures 15.7 and 15.8 are showing the temporal variation in number of cool nights (TN10p) and warm days (TX90p), respectively. It is observed that there is increase in maximum temperature and álso percentage of days when maximum temperature more than 90th percentile also increased in whole basin. It means that warm days showing increasing trend and may further increase in future.

Figure 15.3

Temporal Variation in CDD Index for Five Stations

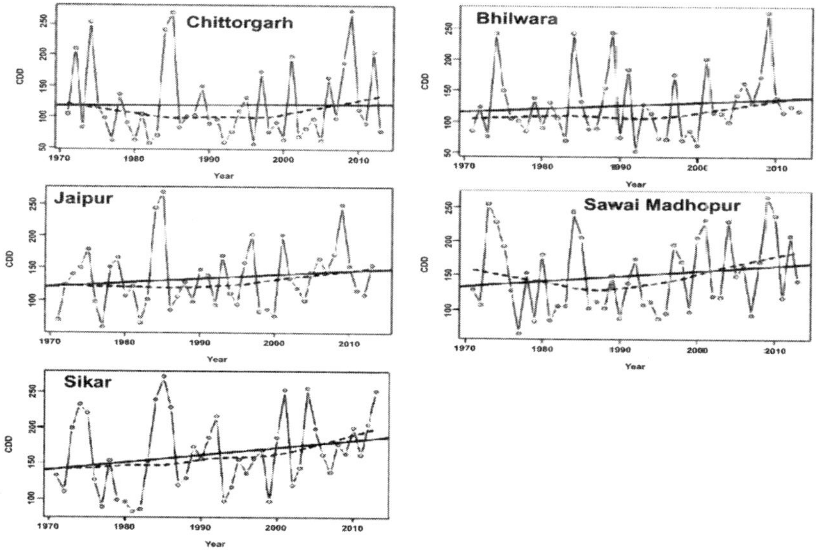

Figure 15.4

Temporal Variation in CWD Index for Five Stations

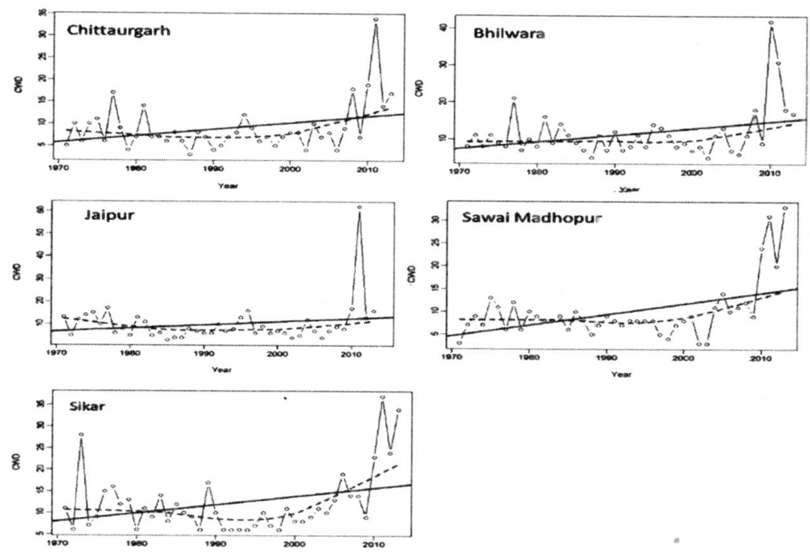

Figure 15.5

Temporal Variation in PRCPTOT Index for Five Stations

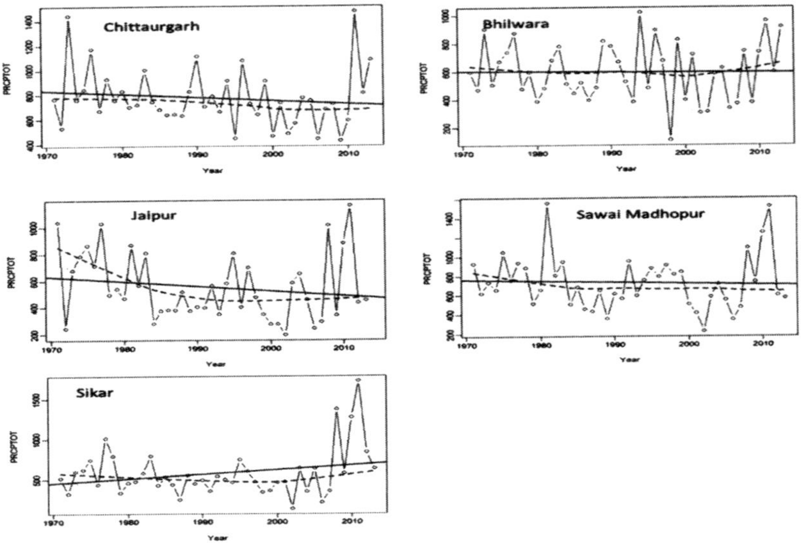

Figure 15.6

Temporal Variation in RX1day Index for Five Stations

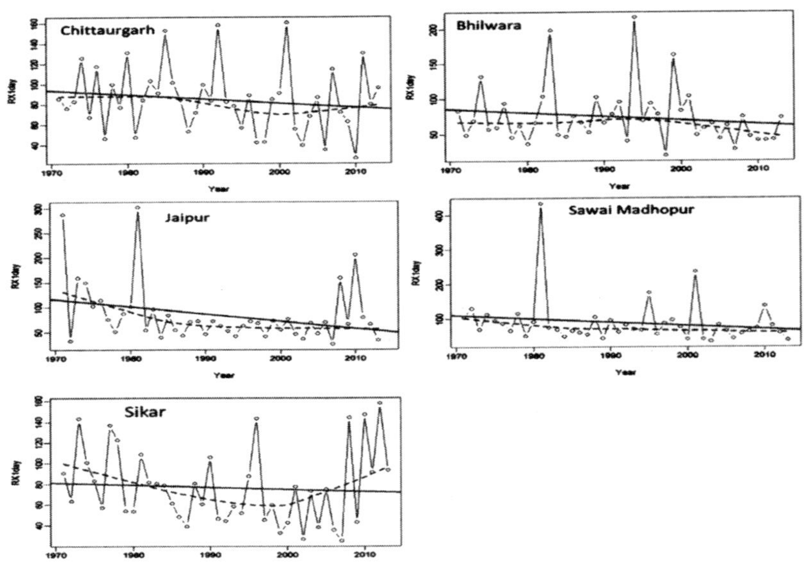

Figure 15.7

Temporal Variation in TN10p Index for Five Stations

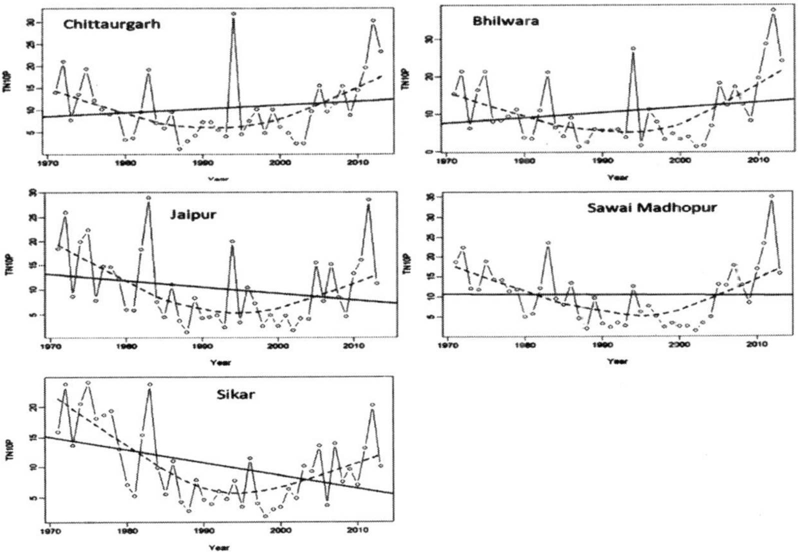

Figure 15.8

Temporal Variation in TX90p Index for Five Stations

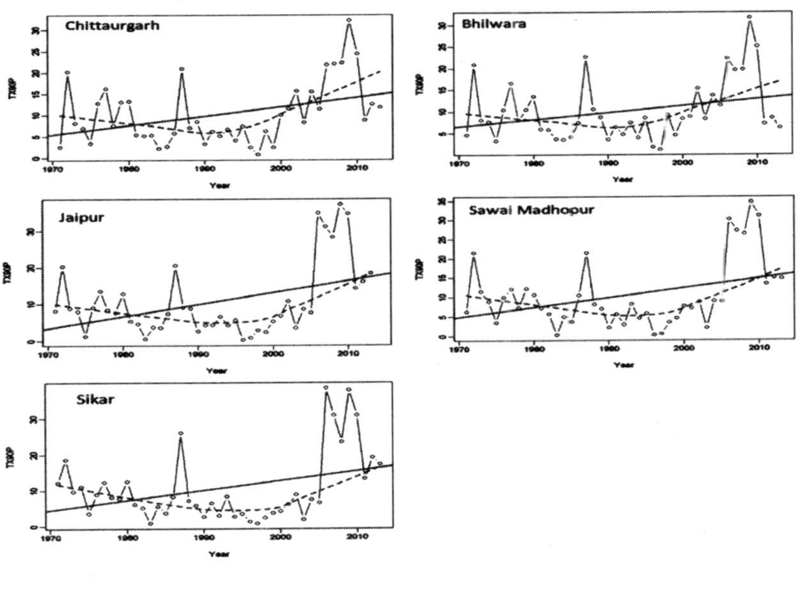

Further analysis is conducted by considering all six indices for three periods, i.e. overall (1972–2013), initial period (1972–1992) and later period (1993–2013). Overall period was divided into two equal periods to see the change in indices at various time scales. Figure 9(a-f) shows the annual average values of all six indices at five meteorological stations for three periods.

Figure 15.9a shows the annual mean CDD values for all five stations for three periods. It was observed that CDD values were higher in the later period (1993–2013) as compared to initial period (1972–2013) for all the stations except Sikar station. The highest value of CDD is observed at Chittorgarh station followed by Bhilwara station. This shows that there is higher number of consecutive dry days in the upper basin with an increasing trend. In Figure 15.9b, there is increasing values of CWD at all stations in the later period when compared with initial period.

In the indices PRCPTOT (Figure 15.9c), it is observed that there is a decrease in total amount of precipitation at all stations except Sikar station. Highest and lowest rainfall is observed at Chittorgarh and Jaipur stations, respectively. Figure 15.9d shows the month maximum 1-day precipitation, higher at the Sawai Madhopur station in the period 1972–2013 and lower at the Jaipur station in the period 1993–2013. It is further observed that there is a decrease in value of RX1Day index at all the stations. As per Figure 15.9e, TN10p (cool nights) is increasing at Chittaurgarh and Bhilwara stations with a sharp decrease at Sikar station when compared for two periods.

In case of TX90p (warm days), there is an increase in value for all the five stations. Highest change was observed at Jaipur and lowest in Bhilwara (Figure 15.9f). The possible reason for this high change in Jaipur is urbanization and urban heat island effect.

Figure 9(a-f)

Average Annual Values of Extreme Temperature and Precipitation Indices for Three Periods

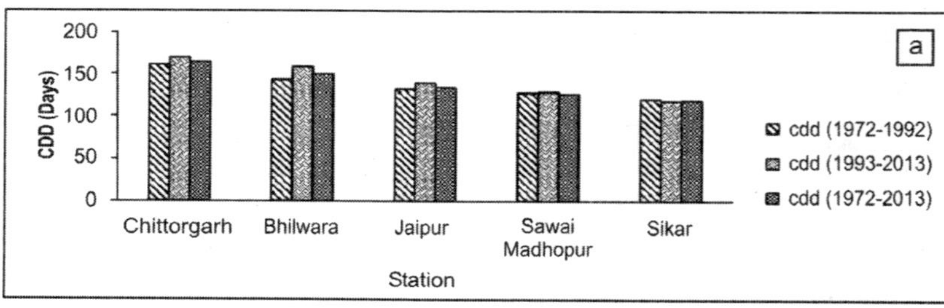

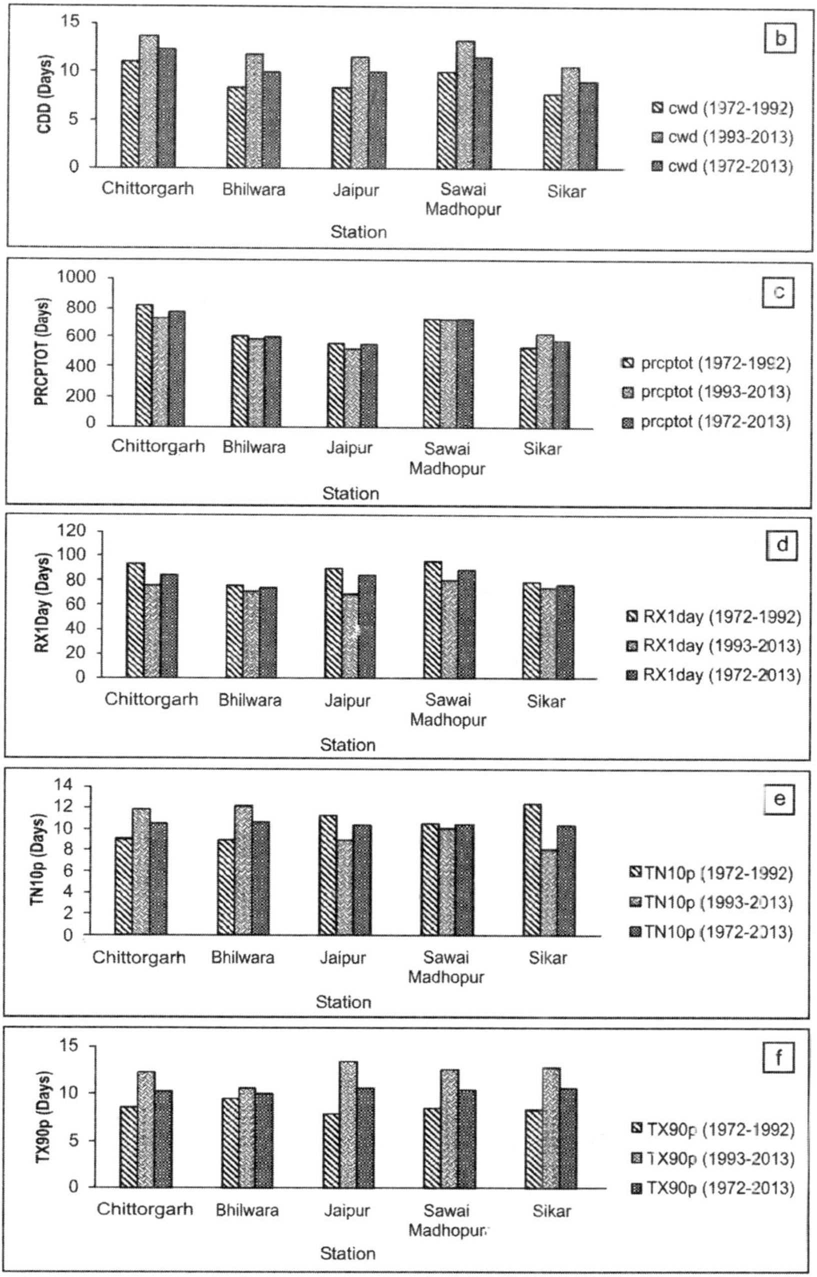
b
CDD (Days)
15
10
5
0
Chittorgarh
Bhilwara
Jaipur
Sawai Madhopur
Sikar
Station
cwd (1972-1992)
cwd (1993-2013)
cwd (1972-2013)
c
PRCPTOT (Days)
1000
800
600
400
200
0
Chittorgarh
Bhilwara
Jaipur
Sawai Madhopur
Sikar
Station
prcptot (1972-1992)
prcptot (1993-2013)
prcptot (1972-2013)
d
RX1Day (Days)
120
100
80
60
40
20
0
Chittorgarh
Bhilwara
Jaipur
Sawai Madhopur
Sikar
Station
RX1day (1972-1992)
RX1day (1993-2013)
RX1day (1972-2013)
e
TN10p (Days)
14
12
10
8
6
4
2
0
Chittorgarh
Bhilwara
Jaipur
Sawai Madhopur
Sikar
Station
TN10p (1972-1992)
TN10p (1993-2013)
TN10p (1972-2013)
f
TX90p (Days)
15
10
5
0
Chittorgarh
Bhilwara
Jaipur
Sawai Madhopur
Sikar
Station
TX90p (1972-1992)
TX90p (1993-2013)
TX90p (1972-2013)

Figure 15.10 shows the pattern in change of the six indices for 5 stations. In CDD, Sikar station shows the negative change (decrease) compared to all other stations. CWD shows the positive change at all the station, which means there will be more number of wet days (rainfall > 1mm). Total annual precipitation is decreasing at all stations except Sikar station. There is negative change at all stations in RX1day which means that there is less rainfall on daily temporal scale. In case of temperature indices, there will be more number of cool nights in the area close to Chittorgarh and Bhilwara stations, but warm days will increase at all the stations as represented by a positive change in TX90p.

Figure 15.10

Annual Mean Change in Precipitation and Temperature Indices

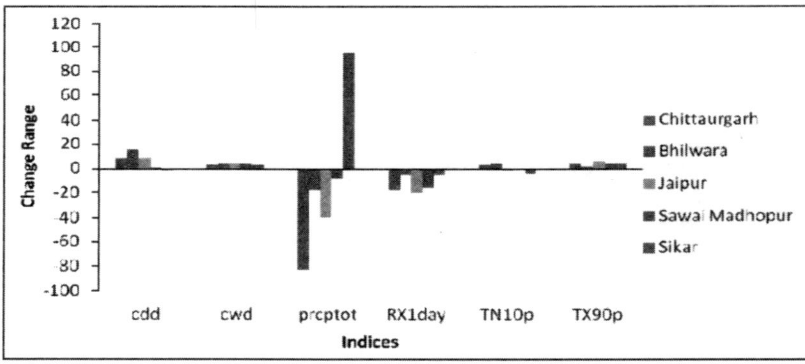

Table 15.4 gives the overall slopes of all the six indices for different periods. Negative means decreasing trend and positive value show the increasing trend in the indices.

Table 15.4

Slope of the Climate Indices for Different Periods

Station	*Period*	*CDD*	*CWD*	*PRCPTOT*	*RX1day*	*TN10p*	*TX90p*
Chittaurgarh	1972–1992	1.77	–0.27	–7.92	0.85	–0.52	–0.31
	1993–2013	3.99	1.14	7.84	0.50	0.57	0.88
	1972–2013	1.05	0.19	–2.76	- 0.42	0.09	0.22
Bhilwara	1972–1992	–1.84	–0.09	–2.63	0.51	–0.55	–0.03
	1993–2013	3.51	0.98	4.68	- 3.19	0.95	0.43
	1972–2013	0.77	0.24	–0.34	- 0.52	0.14	0.09

Cont'd...

Cont'd...

Jaipur	1972–1992	-0.27	-0.30	-15.22	- 2.95	-0.75	-0.26
	1993–2013	1.34	0.72	9.00	1.94	0.49	1.41
	1972–2013	0.60	0.14	-3.63	-1.46	-0.13	0.32
Sawai Madhopur	1972–1992	0.11	-0.14	-11.22	-2.12	-0.66	-0.34
	1993–2013	2.94	0.70	7.59	-1.55	0.93	1.19
	1972–2013	0.52	0.18	-1.34	-1.01	-0.01	0.25
Sikar	1972–1992	-2.07	-0.24	-9.04	-2.32	-0.91	-0.27
	1993–2013	2.86	0.60	29.76	2.24	0.44	1.38
	1972–2013	0.08	0.15	5.78	-0.22	-0.21	0.28

Possible Impacts on Water and Health

Considering the spatial variability of temperature and precipitation (decrease or increase) in the study area, three categories are identified as given in Table 15.5. It is observed that whole basin is under an increase in maximum temperature and major portion of the basin is falling under decreasing precipitation. This shows that basin will suffer from water scarce situation with low surface runoff and less groundwater recharge. There is chance that potential evapotranspiration will increase in the study area. Also, it is well known that reduction in precipitation will enhance the water quality issues in the area. Due to increase in temperature and decrease in precipitation, there is chance of increase in health related problems in the area. Table 15.5 presents the linkage between climate change pattern, water situation and health based on the secondary information in literature and existing studies. The Vulnerability of the society also depends on their sensitivity and adaptive capacity of the human being.

Conclusion

The focus of this study was to assess the climatic change situation in the Banas River Basin using temperature and precipitation data of five meteorological station data. Six climatic indices are prepared for the basin to find out the existing climate change situation in the basin. Different changes are investigated in the different locations and linkage with water and health is identified for different scenarios. The precipitation indices, i.e. PRCPTOT and RX1day, show the negative trends in the whole basin, whereas temperature indicator, TX90p, shows the increase trend which means the warm days are increasing during the period 1972–2013. Based on these identified climatic change conditions, linkage is developed between climatic variables,

Table 15.5

Linkage between Climate Change, Water and Health

Representing Districts	*Precipitation*	*Minimum Temperature*	*Maximum Temperature*	*Possible Impacts on Water*	*Possible Impacts on Health*
Chittorgarh/ Bhilwara	Decrease	Decrease	Increase	Surface runoff- Decrease GW recharge - Decrease PET- Increase Water Quality- Decrease Water available – Reduce	Sunburn; Skin cancer; Asthma; Cholera; Swine flu; Heat Exhaustion; Mental stress; Mortality; Dehydration; Malnutrition; Strain on social relationship
Jaipur/Tonk/ Sawai Madhopur/ Karauli/Dausa	Decrease	Increase	Increase	Surface runoff- Decrease GW recharge - Decrease PET- Increase Water Quality- Decrease Water available – Reduce	Sunburn; Skin cancer; Asthma; Cholera; Swine flu; Heat Exhaustion; Mental stress; Mortality; Dehydration; Malnutrition; Strain on social relationship
Sikar/Nagaur	Increase	Increase	Increase	Surface runoff- Increase GW recharge - Increase PET- Increase Water Quality- Increase Water available – Increase	Dysentery; Sunburn; Skin cancer; Asthma; Cholera; Swine flu, Heat Exhaustion; Mental stress; Mortality; Dehydration; Malnutrition; Strain on social relationship; Food poisoning

Note: GW- Groundwater; PET – Potential Evapotranspiration.

water availability and human health. Due to less precipitation and increase in temperature, more area in the study area will face water scarcity problems. There is a chance that an increase in temperature may increase various health issues related mortality, mental diseases, waterborne diseases and stress on the social life of human beings. This is a more critical situation in water stressed state like Rajasthan and further actions are required in terms of water management, health care centre and proper knowledge of various precautions to public through education and capacity building programmes.

References

Ahmad, I., Tang, D., Wang, T., Wang, M., and Wagan, B. (2015). Precipitation Trends Over Time Using Mann-Kendall and Spearman's Rho Tests in Swat River Basin, Pakistan. *Advances in Meteorology*. 2015.

Akhtar, R. (2007). Climate Change and Health and Heat Wave Mortality in India. *Global Environmental Research,* 11(1): 51.

Alexander, L.V., Zhang, X., Peterson, T.C., Caesar, J., Gleason, B., Klein Tank, A.M.G., and Tagipour, A. (2006). Global Observed Changes in Daily Climate Extremes of Temperature and Precipitation. *Journal of Geophysical Research: Atmospheres,* 111(D5).

Caloiero, T. (2015). Analysis of Rainfall Trend in New Zealand. *Environmental Earth Sciences,* 73(10): 6297–6310.

Del Río, S., Cano-Ortiz, A., Herrero, L., and Penas, A. (2012). Recent Trends in Mean, Maximum and Minimum Air Temperatures Over Spain (1961–2006). *Theoretical and Applied Climatology,* 109(3–4): 605–626.

Hartmann, D.L., Klein Tank, A.M.G., Rusticucci, M., Alexander, L.V., Brönnimann, S., Charabi, Y., Dentener, F.J., Dlugokencky, E.J., Easterling, D.R., Kaplan, A., Soden, B.J., Thorne, P.W., Wild, M., Zhai, P.M. (2013). Observations: Atmosphere and Surface. In: Stocker, T.F., Qin, D., Plattner, G.K., Tignor, M., Allen, S.K., Boschung, J., Nauels, A., Xia, Y., Bex, V., Midgley, P.M. (eds) Climate Change 2013: The Physical Science Basis. Contribution of working group I to the fifth assessment report of the intergovernmental panel on climate change. Cambridge University Press, Cambridge.

IPCC (2007). 'Human Health.' *Fourth Assessment Report.*

IPCC (2014) Climate Change 2014: IPCC Fifth Assessment Synthesis Report-summary for Policymakers. An Assessment of Intergovernmental Panel on Climate Change. Cambridge University Press, Cambridge.

Klok, E.J., and Klein Tank, A.M.G. (2009). Updated and Extended European Dataset of Daily Climate Observations. *International Journal of Climatology,* 29(8): 1182–1191.

Kothawale, D.R., Munot, A.A., and Krishna Kumar, K. (2010). Surface Air Temperature Variability over India during 1901–2007, and its Association with ENSO. *Climate Research.* 42(2): 89–104.

Kovats, S., and Akhtar, R. (2008). Climate, Climate Change and Human Health in Asian Cities. *Environment and Urbanization,* 20(1): 165–175.

Kumar, K.K., Patwardhan, S.K., Kulkarni, A., Kamala, K., Rao, K.K., and Jones, R. (2011). Simulated Projections for Summer Monsoon Climate over India by a High-resolution Regional Climate Model (PRECIS). *Current Science,* 101(3): 312–326.

Modiri E., and Kamali G. (2012). Effects of Climate Change on Catchment Runoff Resulting from Precipitation in Freshwater Chah Nime Lakes in Zahak in 2020, National Conference on Inter-Basin Water Transfer, Shahr kord, Iran. doi: 10.13140/RG.2.1.4576.4885

Modiri, E., and Modiri, S. (2016). Zonation of Vulnerability of Precipitation Changes in the Great Khorasan. *Arabian Journal of Geosciences,* 9(5): 1–11.

Panda, D.K., Mishra, A., Kumar, A., Mandal, K.G., Thakur, A.K., and Srivastava, R.C. (2014). Spatiotemporal Patterns in the Mean and Extreme Temperature Indices of India, 1971–2005. *International Journal of Climatology,* 34(13): 3585–3603.

Parajka, J., Kohnová, S., Bálint, G., Barbuc, M., Borga, M., Claps, P., Cheval, S., Gaume, E., Hlavcová, K., Merz, R., Pfaundler, M., Stancalie, G., Szolgay, J., Blöschl, G. (2010). Seasonal Characteristics of Flood Regimes Across the Alpine–Carpathian Range. *J Hydrol.* 394: 78–89.

Pellicone, G., Caloiero, T., Coletta, V., and Veltri, A. (2014). Phytoclimatic Map of Calabria (Southern Italy). *Journal of Maps,* 10(1): 109–113.

Ray, K., Manoroma, M. and Chinchalikar, J.R. (2009). Climate Variability Over Gujrat, Indiak, ISPRS Archives XXXVIII-8/W3 Workshop Proceedings: Impact of Climate Change on Agriculture.

Sen Roy, S., and Balling, R.C. (2004). Trends in Extreme Daily Precipitation Indices in India. *International Journal of Climatology,* 24(4): 457–466.

Sharma, D., and Babel, M.S. (2014). Trends in Extreme Rainfall and Temperature Indices in the Western Thailand. *International Journal of Climatology.* 34(7): 2393–2407.

Toreti, A., and Desiato, F. (2008). Temperature Trend over Italy from 1961 to 2004. *Theoretical and Applied Climatology.* 91(1–4): 51–58.

Vose, R.S., Easterling, D.R., and Gleason, B. (2005). Maximum and Minimum Temperature Trends for the Globe: An Update Through 2004. Geophysical Research Letters. 32(23).

Vyas, P., Patel, H.R., and Kar, B.I. ISPRS Archives XXXVIII-8/W3 Workshop Proceedings: Impact of Climate Change on Agriculture 118 Impact Analysis of Climate Change on Different Crops in Gujarat, India.

WMO (1996), Climate and Urban Development, WMO No 844, World Meteorological Organisation, Geneva.

Young, S., Balluz, L., and Malilay, J. (2004). Natural and Technologic Hazardous Material Releases During and After Natural Disasters: A Review. *Science of the Total Environment*, 322(1): 3–20.

16

Ground Water Pollution Scenario, Health Issues and Management Options in Rajasthan

L.N. Mathur

Water is a universal solvent because of its extraordinary ability to dissolve a broad range of substances. Most of the arid regions of the world, including Rajasthan, witness naturally occurring inferior quality of ground water. The arid and semi-arid areas with high degree of salinity and high fluoride contents witness the ability of water to dissolve rock/soil material. Rapid increase in population, urbanization, industrialization, enhanced use of fertilizers and pesticides in agriculture, declining ground water levels etc. have also threatened quality of ground water and demands for urgent attention of scientists, administrators and planners. Availability of safe drinking water is a basic human right. However, chemical and bacteriological pollution of drinking water sources pose serious threat to public health in various parts of the State. High salinity, nitrate and fluoride contents beyond permissible limits make the ground water unsuitable for drinking purposes. The aridity increases westward resulting in development of Thar Desert and also Salt Lakes viz. Sambhar, Pachpadra, Phalodi, Didwana, Kuchaman, Nawa etc. Major parts of the western Rajasthan and also parts of Bharatpur, Bhilwara, Jaipur and Ajmer districts of eastern Rajasthan suffer salinity hazards. The arid and semi-arid climatic conditions have also resulted to high fluoride contents in ground water in most of the districts of the State. Ground water is generally alkaline in nature. High nitrate concentration is found in high intensity cropping areas (excess use of fertilizers) and urban areas (poor sanitation and solid waste management). Pollution of ground water by industries and urbanization including at Jaipur, Pali, Jodhpur, Udaipur, Balotra (Barmer district) are other emerging issues of concern in the State.

High Salinity in Ground Water

The analytical data of exploratory drilling indicate that chemical quality of ground water varies from fresh in the southern and eastern parts to highly saline in the western and central parts of the State (Figure 16.1). Electrical conductivity of ground water ranges from less than 250 to over 10,000 micromhos/cm at 25°C in the western and central parts. The saline waters are mostly of sodium chloride type. In a major part of the area in western Rajasthan, the shallow ground water shows electrical conductivity exceeding 5,000 micromhos/cm at 25°C. Such saline tracts are also met within the districts of Bharatpur, Ajmer, Bhilwara and Jaipur in eastern and Nagaur in central parts of Rajasthan State. High salinity is generally confined to aquifers in alluvium, blown sand and the semi-consolidated formations encompassing Tertiary and Mesozoic formations and to a lesser extent in the aquifers of Marwar Supergroup of rocks. In general, ground water quality deteriorates with depth but locally the quality improves vertically and in a few cases depthwise quality of water is highly erratic.

Causes of Salinity

Reasons for enhanced salinity in ground water are many. The Marwar Super Group of rocks as well as Tertiary formations, due to their marine nature/arid deposits, mostly yields brackish to saline ground waters, whereas aridity contributes salinity mainly to Quaternary aquifers. However, ultimate source of salt and origin of salt lakes in Rajasthan is still debatable and leaves plenty of scope for scientific investigations. The main causes of salinity includes:

- Arid climatic conditions resulting in low rains, more evaporation, fine soil texture and sediments, developing inland drainage systems, choking of palaeo-channels and formation of closed basins due to Aeolian activities.
- Neotectonic activities have supported to disorganize the palaeo-drainage systems forming natural depressions and thereby, confluence of several channels with centripetal drainages. Impact of aridity on such accumulated stagnant water has transformed fresh water into saline water.

- Salinity in the area is supplemented by strong south-westerly winds, transporting salts with it.
- Formation of water salinity as in Tertiary, Marwar Supergroup of rocks is due to their marine/arid depositions and connate type of water entrapped in geological past.
- Salinity of ground water in canal command areas is due to prevailing waterlogging conditions.

Quality Variation in Time and Space

There is change in degree of salinity in ground water particularly in canal command areas and also in urban areas. Seepage from canal command areas has improved the chemical quality of ground water at places, whereas waterlogging hazards have increased the soil and water salinity and alkalinity. Studies in Jaipur urban areas have revealed that shrinkage in the open area for ground water recharge and also increasing draft has caused enhanced salinity, particularly in the walled city areas. Deterioration in chemical quality of ground water may be due to overexploitation, urbanization and use of fertilizers, pesticides, etc. in rural areas.

Figure 16.1

Fresh Ground Water Pockets in Western Rajasthan

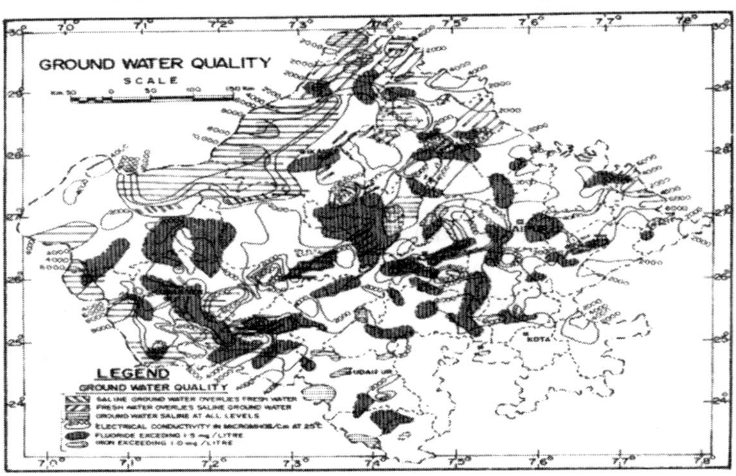

In otherwise saline areas, potable water is found in areas covered by sandstone and limestone of Marwar Super Group, Lathi formation, Tertiary formations and pocket areas in Quaternaries. Thickness of fresh water zone is less than 10m towards the eastern margin of the desert, resulting in ground water drought situation in alluvial aquifers. However, thick fresh water aquifers exist in Jaisalmer district (Lathi basin) and Palaeo-courses of lost Saraswati (around Tanot-Kishangarh-Ghotaru areas), Bikaner, Churu and Nagaur districts (Palana-Nagaur basin), Barmer district (Bhadka area), Jodhpur district (Borunda limestone, Jodhpur sandstone in Teori-Mathania area), etc.

Health Issues

High salinity in water makes it unsafe for drinking purposes, as per the international and national norms/specifications. However, field experiences says that such norms may be partially applicable in arid/arid to semi-arid terrains of Thar Desert. Hot climate causes lot of sweating containing salts with it. Drinking of water of higher salinity can only save the deserters from dehydration. This fact is further confirmed by lesser cases of high blood pressure and related cardiac diseases in the areas. Thus, the nature has already made a balance for arid zone people. However, with the change in lifestyle especially in towns and urban areas, catering for fresh drinking water supply has become need of the hour in arid and semi-arid regions of the State.

Saline Water Management

In view of the occurrence of huge quantity of saline ground water spread over larger areas; it may not be technically feasible/economical viable for in-situ desalination/dilution/ flushing out the salinity. Possibilities for bio-remediation for desalination need to be searched out. Since the State is having huge quantum of saline waters especially in western parts, assessment and research should be taken up for economic desalination, salt production and proper saline water management. Saline water management may include following options:

- Blending of saline ground water with canal water/other surface water sources for drinking purposes wherever situation demands/feasible.
- Growing salt tolerant crops in coarse textured soil where soil salinity is flushed out due to rains or can be improved by adding gypsum etc.
- Cost effective desalination of saline water using solar energy/wind energy.
- Popularization of desalination techniques and people's participation to run and repair such desalination plants is needed for effective and efficient working.

- Exploring possibilities of salt-water fisheries.
- Economical salt extraction from saline waters in selected areas.
- Dual water supply schemes may be initiated. Safe drinking water may be supplied for lesser time and volume, whereas water supply for other domestic purposes (bathing, cleaning, washing, etc.) may be given for longer time span.
- Artificial recharge to ground water flushes and dilutes salinity in ground water and therefore, needs to be encouraged. Large-scale recharge may develop floating fresh water layer that can be exploited during required period.

Fluoride Pollution in Ground Water

Most of the arid and semi-arid regions of the world have shown accumulation of high fluoride contents in ground water. Rajasthan is the driest State of India where high salinity and fluoride content in aquifer water make the scenario still worst. Areas with fluoride exceeding 1.5 mg/l occur in almost all the districts of Rajasthan under diverse hydro-geological conditions. The widespread high concentration of fluoride in the ground water in Rajasthan inter-alia appears to be related to arid/semi-arid present as well as palaeo-climatic conditions. Public Health Engineering Department revealed on the basis of habitation survey (1989–93) that out of 37,889 villages in the State 9,740 have more than 1.5 ppm (parts per million) of fluoride in ground water and 3,280 villages have more than 3.0 ppm of fluoride. Subsequent survey finds 11,909 villages and 11,388 other habitats having fluoride level over 1.5 mg/l. Similarly, 4,250 villages and 3,449 other habitats are having fluoride concentration over 3mg/l. The worst effected districts are Nagaur, Jaipur, Ajmer, Sirohi, Pali, Jhunjhunu, Churu, Tonk, Dungarpur and Udaipur. Bankapatti is the term used for high fluoride areas (*Banka* meaning oblique, *Patti* meaning spinal chord). The Bankapatti exists in a longitudinal belt extending from north of Nagaur to Indawar in Merta block and as few pockets covering parts of Parbatsar, Makrana, Didwana and Ladnu block of Nagaur

In major parts of the Sanganer block area, ground water is being exploited from alluvium but in southern parts of the block area, ground water from granite, gneiss is also exploited. In major part of the block, depth to water level lies in the range of 20 to 25 m. Ground water is saline in the south-west of the Newta canal. Sanganer town is famous for the dyeing, printing and processing industry. The effluent from various factories situated in and around Sanganer town is being discharged into the canal, situated on the Diggi Malpura road and is flowing in the westward direction towards Mohana and Nevta. A branch of the canal also flows in the southward direction towards Watika. Investigations reveal that a majority of the factories are discharging their effluent into the canals, the major part of which flow towards Nevta. A few industries are discharging their effluent into the canal flowing in the southward direction, but the discharge is maximal and the flow in the canal is very low.

Effluent being discharged into the canal is flowing towards Mohana and then towards Nevta, where it gets accumulated on account of a bandh, which was limited for impounding rainwater. High fluoride value obtained in the canal water flowing from Sanganer towards Nevta might be on account of fluoride in the raw water being used by the industry. The dyes being used by the industry also contain high fluoride and where huge quantity of such dyes is discharged as effluent, the water in the canal gets enriched with fluoride. Extensive irrigation practices from such high fluoride canal/nala water spreads and enrich the fluoride contents in ground water. The non-industrial areas in the Sanganer block and also adjoining block areas are also fluoride affected. Thus, it is inferred that high fluoride problem in Sanganer block area is natural as well as anthropogenic in origin.

Observations about Impact of High Fluoride

- Alkalinity of drinking water, calcium, Vitamin C along with molybdenum, aluminum and phosphates intake determine the impact of fluoride on human body causing fluorosis.
- Fish-food, tea, cottonseed and some wines may have relatively larger amount of fluoride. Drugs used for treatment of osteoporosis and related metabolic disorders, certain anesthetic drugs, fluoridated corticosteroids and dentine obludents are said to contain fluoride, which can be metabolized in the body.
- In areas, where water is hard due to calcium and magnesium contents, the prevalence of skeletal fluorosis is relatively less. In western parts of the State, milk and milk products containing high calcium is the main food reducing the adverse impact of high fluoride in drinking water.
- A more severe type of mottling has been observed in children who had a low intake of Vitamin C. Citrus fruits are reported to reduce the toxic effect of fluoride. Thus, consumption of fresh fruits is recommended in fluorotic belts.
- With increase in age, prevalence of dental mottling and subsequently skeletal fluorosis increases in the fluorosis belts. By the

end of the third decade the prevalence reaches to a maximum and plateaus thereafter. Although dental fluorosis is observed in the deciduous teeth but its prevalence starts rising from the time when permanent teeth appear. In highly endemic zones, people become edentulous in the fourth and fifth decade of life and large number of them develop hypnosis.

- In fluorotic zones, females suffer less than males due to their migration after marriage and higher consumption of drinking water by man doing strenuous physical work. But women are more affected during pregnancy and breast-feeding causing lack of sufficient calcium in their body.
- People involved in mining fluoride rich ores and working in aluminum industries are more likely to suffer from fluorosis as witnessed in Udaipur, Dungarpur and Bhilwara districts.

Testing for Fluorosis Patients

The three commonly used tests to assess the presence of skeleton fluorosis include coin test, chin test and hand stretch test, which can be carried out anywhere in the field. Test includes asking the individual to touch the chest with the chin. While a normal person can easily do it, a fluorosis patient is unable to bend the neck. Thus touching the chest with the chin is not possible. In the coin test, the person is asked to lift a coin from the floor without bending the knees. A fluorosis patient is unable to lift the coin from the floor without folding his knees. A normal healthy individual can stretch the hands, fold the arms and touch the back of the head effortlessly. But fluorosis affected person is unable to stretch the hands, fold the arms and touch the back of the head.

Fluoride Content in Food Items

Fluoride rich drinking water is generally the major source of fluoride entering the body. But food items can also contribute significantly to the total fluoride intake. The significance of this observation is that fluoride absorption from food is 50 per cent to 80 per cent in contrast to essentially complete absorption from drinking water. Many food items are a rich source of fluoride. The dried sea weeds, tea, sardines, shrimp etc. are high fluoride containing food.

Health Hazards

Fluorine is one of the essential elements in our body. It is absorbed through gastrointestinal tract (except insoluble organic fluoride compounds) and is bound in the mineralizing tissues or is distributed in the soft tissues, tissue fluid and serum. The inorganic fluoride in blood is quickly removed by the kidneys. In infants, small amount of fluoride is beneficial as it inhibits caries attack and induces the saliva to precipitate salt that heal mild pitting on the enamel surfaces. Fluoride levels of 1 mg/litre in drinking water reduce dental cavities in children. The World Health Organisation (WHO) recommended fluoride contents in drinking water as 1.0 to 1.2 mg/l in temperate and 0.5 mg/l in tropical climates. Excessive fluoride may accumulate in bones and give rise to clinical manifestations i.e. fluorosis. Its very high intake daily produces dental fluorosis, Osteo-fluorosis and damage to kidneys. Rates of fluorosis decline if the source of fluoride is removed. Fluorides have an inhibiting effect on enzymes. Persons under malnutrition (lack of protein, Vitamins C and D) are more prone to fluorosis. The first symptom of dental fluorosis is development of opacities followed by brownish staining and pitting. Infant during enamel forming period is the most vulnerable and maleruption of permanent teeth, double teeth, abnormal inter-dental distances etc. have also been observed due to high intake of fluoride. Chronic excessive fluoride intake may leads to skeletal fluorosis (osteo-fluorosis), i.e. hardening or abnormal density, crippling, genu valgum knock knee, etc. Fluoride interference with enzymes is responsible for bone degradation and deposition of excessive bone tissues, which causes difficulties in bending of joints/spinal column and painful neurological symptoms appear when nerves of spinal cord are compressed.

Other Ill-Effects

- Red blood cells are also affected by accumulation of fluoride on their membranes, which in turn loose calcium content. The RBC die and patient suffers from anaemia.
- Gastrointestinal Mucosa is involved in the early stage of fluorosis. The main diseases are nausea, loss of appetite, stomach pain, gas formation, bloated feeling, constipation with intermittent diarrhea and headache.
- Sperm abnormality – oligospermia (deficiency of sperms in semen), azoospermia (absence of sperms in semen) and low testosterone levels in blood is very common.
- Neurological manifestation like nervousness, depression/ tingling sensation in fingers and toes, excessive thirst and feeling for frequent urinate, etc.
- Other adverse effects viz. dying of cattle, corrosion of utensils, etc.

Nitrate Pollution in Ground Water: Jaipur Urban Area

Rapid pace of urbanization has overtaken the development of sanitation and waste disposal facilities resulting in high nitrate pollution in ground water. Maximum permissible limit of nitrate for drinking water is 100 ppm (BIS norms). Nitrate concentration beyond this permissible limit in ground water is considered as nitrate pollution. Quaternary formations are the principal aquifers in Jaipur urban agglomerate. These are porous formations, which provide quick opportunities to pollute the ground water. Depth to water level varies from 8 m to over 50 m. Walled city area witness shallower water level which further facilitate its early contamination. Ground water, by and large, flows towards southerly and south-westerly direction.

Source of Nitrate Pollution

1. Leakage and seepage from sewerage system and soak pits is the chief source of nitrate contaminating ground water.
2. Lack of solid waste disposal management is resulting in nitrate pollution in ground water.
3. Unlined drains, which carry effluents and sewage passing through the city are sources of high nitrate in ground water in the walled city and adjoining areas.
4. Cattle excreta may contribute to nitrate pollution locally.
5. Use of excessive nitrate fertilizers for gardening, etc. may attribute to high nitrates locally.

Affected Areas

Wall city area is the worst affected area, which extends from Chand Pole to Jorawar Singh Gate and Suraj Pole. Similarly Jal Mahal Lake, Gaitor and Kala Hanuman Mandir area are also severely affected. Colonies adjacent to wall city including C-Scheme, Bani Park, Civil Lines, Jyoti Nagar, Adarsh Nagar, Tilak Nagar, Sikar House, parts of Shastri Nagar, etc. are also affected areas of pollution.

Health Hazards

Excessive intake of drinking water with nitrate content beyond permissible limit (100 ppm) may cause clinical manifestations of disease called Methemoglobinemia (blue baby) in infants. Signs of the disease include listlessness and a bluish tinge to the skin. In adults, intake of high nitrate water may cause stomach disorder or gastrointestinal diseases. Cattle are also susceptible to high nitrate water and may cause loss of milk production and birth to aborted calves.

Management Options

- Proper sewerage system should be developed and all the houses be connected with it.
- Soak pits should be closed and banned.
- All the unlined drains carrying sewerage/effluents needs to be lined.
- Use of nitrate fertilizers for gardening, etc. should be discouraged.
- Cattle establishments should be shifted out of the urban areas.
- Plantation should be encouraged, as they will utilize the excess nitrate as fertilizers.
- Artificial recharge schemes needs to be implemented wherever feasible. This will dilute and flush the polluted ground water.
- All the ground water abstraction structures including hand pumps with high nitrate concentration should be marked by red paint so as to avoid their utility by the common people for drinking purposes.
- All the wells yielding high nitrate water should be closed immediately by the Public Health Engineering Department.

Industrial Pollution: A Few Studies by CGWB

Jodhpur Industrial Area

During the last few decades, group of textile, chemical, metallurgical, ore processing and plastic industries have come into existence. The city sewerage and effluents from the Basani industrial area are dumped into the ephemeral Jojari River. The geological formation occupying the area is aeolian sand and alluvium. Alluvium is underlained by Jodhpur sandstone and shale in the northern part of the area and by Malani igneous rocks in southern part. Hydrogeological formations occurring in the area are Malani granites and Quaternary Alluvium. Quaternary formation forms aquifer in the northern part of the area. Depth to water level varies from 3 to 50m. Unplanned disposal of the effluents coming out of the group of industries and also the city sewage affect the ground water sources of the area covering Jhalamond, Tanawar, Salawas and Nanadawan villages lying along the river Jojari. The industrial effluents and the city sewage are being finally discharged into this river. The textile and Stainless Steel Rolling Mills generate about 13.5 million litres per day (MLD) to 16.7 MLD liquid effluents per day, which is discharged into the nearby ephemeral streams polluting ground water resources. The effluents so generated are coloured, turbid along with high COD and BOD and characterized by salinity. There are several places like Jhalamand, Kudi, Sangaria, Tanavara along the riverbed, where sulphate levels are more than 400 ppm causing health hazard. The concentration of NO_3 in ground water varies from 11 ppm to 215 ppm. The high concentration of nitrate in local patches reveals the possibilities of contribution by leaching of excess of nitrate fertilizer, raw effluents consisting of salts of nitrate and sewerage effluents. Fluoride content in ground water in the area is normally less than 2.0 mg/l. High concentration of fluoride have been observed at Sangaria, Kuri, Mogra, Tanavara and Salawas.

Khetri Copper Area: Jhunjhunu District

Khetri Copper Complex is one of the important mining activities in the Jhunjhunu district of Rajasthan. Water/effluent samples collected from the area were analysed for basic constituents and heavy metals including Zn, Cu, Ni, Cr, Cd, Mn, Ce and Pb. The results of analysis reveal that all the constituents in ground water are within the permissible limit. However, the high levels of Total Hardness may be because of dumping of residues in the nearby valley.

Kota Industrial Area

Kota city is known for thermal power station as well as industries. The results of studies reveal that electrical conductivity in the area ranges from 430 micromhos/cm to 5,900 micromhos/cm (Raipura Nalla industrial area). Chloride concentration in ground water varies from 28 ppm to 1,030 ppm. Nitrate content in ground water is generally within permissible limit of 100 ppm. Fluoride content in ground water is also generally within the permissible limit of 1.5 mg/l except at places like Raipur where it is 3.5 mg/l. High concentration of iron has also been observed.

Udaipur District: Urban Area

Three industrial areas namely Sukher, Mewar Industrial Area (MIA) and Zinc Smelter, Debari were studied. In Sukher Industrial area, most of the industries are marble based, disposes its effluent after treatment into a kaccha drain; the same gets mixed with effluent coming from marble cutting and polishing industries. The drain ultimately leads to a surface barrier constructed west of Brahmanon Ka Gudha. Ground water shows contamination all along the Nalla, which ultimately meets Ahar. In Mewar industrial area, mineral grinders own most of the industrial plots. Major pesticide producing industry is also located in the area. The farmers in the field mostly use the treated effluent from distillery and residue is drained into the Ahar River. Ground water around industrial area is contaminated. Higher value of chromium and cadmium is reported in the area.

Zinc Smelter: Debari

Most of the samples collected in the villages around Hindustan Zinc Limited, the effluent drain shows electrical conductivity varying between 4,000 and 7,000 micromhos/cm with SO_4 values ranging 415 to 2,200 ppm, fluoride between 0.62 to 5.26 mg/l. The wells in Bichri, Thoriya Mayri, Dabok, Sihara and Gorla show contamination. Though the value of zinc in all the cases are within the permissible limit, higher values of cadmium, manganese and lead are present.

Pali Industrial Area

Pali is known for cloth dyeing industries discharging effluents from both the banks of the river Bandi. There are red as well as orange categories of industries in Pali. Seepage from industrial effluents are resulting in pollution of ground water and is rarely flushed out through occasional rainwater runoff in Bandi river. Downstream of Bandi River, near Nera village, a dam has been constructed. The reservoir of the dam is affected by polluted water. Irrigation by such water polluted by dyeing industries is spoiling the agriculture fields further downstream of the dam. The analyses of industrial effluent and dug wells show high concentration of TDS, chloride and sulphate and high electrical conductivity. Lead content in some of the samples is above permissible limit of 0.1 mg/l. BOD and COD of untreated wastewater has been recorded as 43 to 140 mg/l (permissible limit 30 mg/l) and 464 to 1,320 mg/l (permissible limit 250 mg/l) respectively. Effluent treatment was not able to make the water colourless and reduce TDS. Treated water is not fit for irrigation due to high content of TDS and can only be reused in dyeing industries. There are inadequate treatment plants, which is not sufficient/required standard to treat the total volume of wastewater generated by industries.

Balotra Textile Industries

Balotra is the second largest town of Barmer district and developed as industrial centre for textile processing. The textile processing industries generate industrial waste effluents, which contains toxic elements. The industrial effluent is left in river Luni causing ground water pollution downstream of the river and if used for irrigation purposes it will have a damaging impact on the fields. Common Effluent Treatment Plants (CETPs) at Balotra, Jasol and Bithuja, are not able to handle the entire wastewater received and a significant portion was being discharged directly into the river Luni, bypassing the treatment system.

Industrial Pollution Management

All the effluent disposal channels in the affected areas must be appropriately lined, interconnected and brought to common disposal points and Common Effluent Treatment Plants on cooperative basis or government run plants may be installed to effectively treat industrial effluents before their disposal. Remedial measures may be taken for all the industrial areas including Jodhpur, Pali, Balotra, Udaipur, Kota, Khetri (Jhunjhunu district), Bhiwari (Alwar district), Alwar industrial area etc. Large scale artificial recharge to ground water wherever feasible with extreme care may dilute the problem.

Discussion

Water quality management has attained paramount importance worldwide in view of growing concerns on environment and health related impacts. In India, National Water Policy 2002 has given due importance to water quality emphasizing developing innovative techniques based on scientific studies in order to protect our water resources. The policy adopts principle of 'polluter pays' and formulation of necessary legislation. The policy suggests intensive research efforts for water quality by taking up R&D schemes for developing cost effective techniques for pollution abatement, especially for removal of arsenic, fluoride and other toxic substances, identification of sites for dumping toxic waste material to prevent its movement to the ground water, encourage appropriate use of fertilizers, pesticides and insecticides. Water quality aspects have also been mentioned in one of the targets of the Goal–7 'Ensure Environment Sustainability' in the United Nation Millennium Development Goals. National Green Tribunal is taking a serious view of pollution especially in Balotra, Jodhpur and at Pali industrial areas in respect of State of Rajasthan and has banned/given appropriate directions so as to address environmental issues.

The existing arid and semi-arid climatic conditions in the state of Rajasthan has resulted in high salinity and fluoride concentration in ground water whereas industries, urbanization along with excessive use of fertilizers/pesticides are the major man-made pollution in the State. Further, global climate change in the times to come may have adverse impact on quality of ground water and warrants especial attention of the water managers. Geogenic as well as anthropogenic pollution is a serious issue due to their concern with public health in the State. Fresh ground water resources are gradually shrinking owing to enhancing pollution in the State. Gravity of situation, therefore, calls for addressing various pollution-related issues appropriately and effectively from policy planners at Central and State levels to percolating down to grass root levels. R&D activities on pollution related issues needs to be intensified manyfold for suggesting cost-effective practical solutions. Legislative interventions as per Environment (Protection) Act, 1986 and other such Central/State legislations need to be strictly and effectively enforced. Pollution abatement interventions are not only essentially required for addressing public health but also for improving environment and socio-economic fabric of the State and development of the country as a whole.

References

Central Ground Water Board, Groundwater Conditions and Management options, Jaipur Urban Area, Rajasthan, Unpublished Report.

Environment (Protection) Act, 1986.

Mathur, L.N. (2011). Managing Sambhar Salt Lake in Semi-arid Region of Rajasthan, West India, 14th World Lake Conference at Texas, USA.

Mathur, L.N. (2005). Indications of Arsenic Pollution in Ground Water in Rajasthan, Seminar on Water Scenario, Efficient use and Management in Rajasthan Organized by Central Ground Water Board.

Mathur, L.N. (2005). Management Strategies for Nitrate Pollution in Ground Water, Jaipur Urban Area, Rajasthan, India, National Conference on Environmental Conservation organized by BITS, Pilani.

Mathur, L.N. (2005). Saline Water Management in Arid and Semi-arid Zones of Rajasthan, 92nd Indian Science Congress–2005 at Nirma University, Ahmedabad.

Mathur, L.N. Urbanization and Environmental Pollution in Groundwater, Jaipur Urban Area, Rajasthan Organized by Subodh College, Jaipur.

Contributors

A.B. Gupta, Professor, Department of Civil Engineering, Malviya National Institute of Technology, Jaipur.

A.N. Arora, Professor, Department of Civil Engineering, Kautilya Institute of Technology and Engineering College, Jaipur.

Alpana Kateja, Professor, Department of Economics, University of Rajasthan, Jaipur.

Anshul Singh, Research Scholar, Department of Chemistry, Kurukshetra University, Kurukshetra.

Ashu Chaudhary, Assistant Professor, Department of Chemistry, Kurukshetra University, Kurukshetra.

Chitra Choudhary, Assistant Professor, Department of Economics, University of Rajasthan, Jaipur.

Devesh Sharma, Assistant Professor, Department of Environmental Science, Central University of Rajasthan, Ajmer.

Jaishree, Project Fellow, UPE Study of Ecology and Economy of Rajasthan, University of Rajasthan, Jaipur.

Kanika Saxena, Research Scholar, Department of Civil Engineering, Malviya National Institute of Technology, Jaipur.

L.N. Mathur, Scientist, Central Ground Water Board, New Delhi.

Mukta Agrawal, Associate Professor, Department of Home Science, University of Rajasthan, Jaipur.

Neelam Kothari, B.Tech, Department of Civil Engineering, Malviya National Institute of Technology, Jaipur.

Nisha Jain, UGC Post-Doctoral Fellow, Environmental Toxicology Unit, Department of Zoology, University of Rajasthan, Jaipur.

P.K. Parchure, Regional Director, Central Ground Water Board, Western Region, Jaipur.

Pradeep Bhatnagar, Professor and Dean, Faculty of Science, IIS University, Jaipur.

Priyanka Sharma, Lecturer, Department of Zoology, Agarwal PG College, Jaipur.

Ritu Babel, Research Scholar, Department of Home Science, University of Rajasthan, Jaipur.

Rohit Goyal, Professor, Department of Civil Engineering, Malviya National Institute of Technology, Jaipur.

Sanjukta Saha, Assistant Professor, Department of Economics, University of Rajasthan, Jaipur.

S.C. Joshi, Professor, Department of Zoology, University of Rajasthan, Jaipur.

Shruti Jain, M.Tech, Department of Civil Engineering, Malviya National Institute of Technology, Jaipur.

Sreemoyee Chatterjee, Associate Professor, Department of Biotechnology, IIS University, Jaipur.

Sunil Kumar Gupta, Consultant, Pediatrician and Environmental Scientist, Krishna Ram Hospital and Research Centre, Jaipur.

Swatantra Kumar Dubey, PhD Research Scholar, Department of Environmental Science, Central University of Rajasthan, Ajmer.

Yasmeen Khan, Project Fellow, UPE Study of Ecology and Economy of Rajasthan, University of Rajasthan, Jaipur.